Solid Modeling with Pro/ENGINEER™

Brett H. Hamlin

Michigan Technological University

Sheryl A. Sorby

Michigan Technological University

Prentice Hall

Prentice Hall
Upper Saddle River, NJ 07458

Library of Congress Cataloging-in-Publication Data

Hamlin, Brett H.
Solid modeling with Pro/ENGINEER / Brett H Hamlin, Sheryl A. Sorby
p. cm.
ISBN 0-13-490178-9
1. Engineering graphics. 2. Pro/ENGINEER 3. Engineering models--Data processing.
I. Sorby, Sheryl Ann, 1959- II. Title.

T353.H234 2000
620'001'13--dc21
99-088971

Acquisitions Editor: *Eric Svendsen*
Editorial Assistant: *Kristen Blanco*
Vice-president and director of production and manufacturing, ESM: *David W. Riccardi*
Vice-president and editorial director of ECS: *Marcia Horton*
Executive Managing Editor: *Vince O'Brien*
Managing Editor: *David A. George*
Editorial/production supervision: *Lakshmi Balasubramanian*
Manufacturing Buyer: *Pat Brown*
Marketing manager: *Danny Hoyt*

Prentice Hall

Prentice-Hall, Inc.
Upper Saddle River, New Jersey 07458

Printed in the United States of America
10 9 8 7 6 5 4 3 2 1

ISBN: 0-13-490178-9

Prentice-Hall International (UK) Limited, *London*
Prentice-Hall of Australia Pty. Limited, *Sydney*
Prentice-Hall Canada Inc., *Toronto*
Prentice-Hall Hispanoamericana, S.A., *Mexico*
Prentice-Hall of India Private Limited, *New Delhi*
Prentice-Hall of Japan, Inc., *Tokyo*
Pearson Education Asia. Pte. Ltd., *Singapore*
Editora Prentice-Hall do Brasil, Ltda., *Rio de Janeiro*

This book is dedicated to Henry and Fred. May the discoveries of life continue to flow. B.H.H

To all the tough, strong women who helped shape me: Flora, Ethel, Irene, Mariam, Carrie, and Claire. S.A.S

Contents

Preface

The intent of this text is to teach the reader about constraint-based solid modeling by focusing on the functionality and use of a single software package: Pro/ENGINEER. The text was written using Pro/ENGINEER Version 20; however, because its user interface is relatively stable, I believe that the book is more than adequate for use with Pro/ENGINEER 2000. There are subtle differences incorporated into the newer version of the software but most changes are relatively self-explanatory.

The first chapter is meant to be a general introduction to design in the context of computer tools and concurrent engineering. The second chapter explains the user interface and file management system for Pro/ENGINEER. Students do get opportunities in this chapter to "get their feet wet," by following the Guided Tours. Chapter 2 is also meant to serve as a resource. The third chapters covers the basics of two dimensional constraints. Once constraints are learned in 2-D they are more easily transferred to 3-D. The forth chapter applies the skills learned from Chapter 3 and extends them to three dimensions. In Chapter 4 you will find topics of extrusions, revolves, and helical sweeps. The fifth chapter, builds upon parts of Chapter 2, but within the context of being able to create intricate three dimensional objects. Chapter five goes into a mode advance discussion on displaying 3-D parts, this includes perspective views. The sixth chapter discuss some strategies for construction of more complex three dimensional models. Again, this chapter builds upon the previous chapters. Chapter 7 introduces advance modeling techniques which include sweeps and blending of various cross sectional geometry. The final chapter, Eight, covers the methods involved in turning your 3-D model into a working engineering drawing.

When I teach my course in engineering graphics, I usually tell the students two things on the first day of class. First, and foremost, I inform them that I will not be teaching them how to use the software. The only way that they can *learn* the software is by sitting in front of the computer and *using* the software. I will serve as their "tour guide" through the software. I can show them shortcuts and help them focus on certain techniques, but I cannot *teach* them how to use the software—they must do that by themselves. Second, I tell them that if they have previous experience with a 2-D drafting package, such as AutoCAD, they should check that knowledge at the door. I have seen many people struggle to learn Pro/ENGINEER because they try to *force* the software to act like a 2-D drafting package. I try to illustrate the difference between 2-D CAD and 3-D modeling by asking students to imagine the different approaches they would take if they were going to draw me a picture of a table versus if they were going to build me a table out of plywood, nails, a frame, and legs. Most students can easily see that drawing a picture of a table is inherently different than building the table itself. This text is not a "cook book." It is design for more discovery based learning.

The exercises in each chapter are meant to reinforce topics from the chapter; however, in many cases, exercises could be used interchangeably between chapters. In the course where I use Pro/ENGINEER, I usually assign at lease half of all the problems at the end of each chapter. Another good source for homework assignments is engineering graphics texts.

Acknowledgments

I would like to thank Sheryl Sorby for all the support and guidance she has given to this project, she is not only a co-author but a friend. I must also acknowledge all the ME104 students who had to test drive the various iterations of the text. I would like to thank Eric Svendsen, David George, and Lakshmi Balasubramanian of Prentice Hall for seeing this project through to the bitter end. And last, but not least, I would like to thank my wife, Amy Jo, for putting up with my long hours in the office, frozen pizza dinners, and all the projects around the house I didn't finish as I worked on the text.

CHAPTER 1

Engineering Design

What distinguishes engineers from scientists is that engineers solve problems and "make" things. Scientists study problems and report their findings in the literature, whereas engineers try to solve the many technical problems of the world. For example, environmental scientists studied the Earth's atmosphere several years ago and found that as a world society we were creating too many air pollutants and a significant part of the problem stemmed from automobile emissions. Engineers acted on the results of these studies and designed cars which were more fuel efficient and polluted less. Other engineers worked on changing the composition of the fuel so that fewer pollutants were produced during combustion. Engineers are now working on solar-powered or electrically powered vehicles to further reduce the quantity of air pollutants emitted into the atmosphere.

Engineering is a design-oriented profession. We design devices or systems, figure out how to mass-produce them, determine how to package them, and develop ways to ship them to their destination. As an engineer, the type of device or system you design may depend on your discipline, but the design procedure you use is essentially the same regardless of your industry. In this text, you will focus on mechanical design in general and specifically on computer-aided design.

1.1 The Design Procedure

Design is a multistep process, however, there is considerable disagreement on exactly how many "steps" are involved. In this section, we will not try to delineate the different design steps, nor will we try to determine the exact number of steps in the process. We will focus on design as a process rather than a series of steps.

The most important thing you need to understand about the design process is that it is iterative. You have probably heard about the Wright brothers' historic first flight at Kitty Hawk. What you may not be aware of is that they worked on many designs over several years until they finally developed the design solution that enabled them to make their first flight. The history books rarely tell you about the years of frustration and trial-and-error designs that the Wright brothers tested—they only tell you of the end result.

When designing a device or a system, there are certain factors that the engineer must keep in mind. The factors are usually related to the function and cost of the resulting system. In most cases, engineers must make choices during the design process. Consider the automobile, for example. Engineers may choose from metals, composites or plastics for the materials in the car body. Each material type has different advantages and disadvantages for use in this application. Steel is strong and ductile but is prone to corrosion (rust), especially in areas where road salt is used during the winter months. Composites are strong and able to absorb a great deal of energy during crashes, but they can be brittle, and they may be more expensive than steel. Plastics are readily formed into almost any shape and are corrosion resistant, but they are relatively weak, making safety a significant concern. Although plastics are widely used in car bumper systems, they are typically not considered suitable for the car body itself. These are just a few of the factors engineers must consider during the design of an automobile body. Engineers make choices by weighing the often conflicting factors associated with function and cost. A car could be built which would not only be safe but also corrosion resistant, however, the average person may not be able to afford it. Consequently, engineers make trade-offs between cost and safety to design a car which is reasonably safe and relatively inexpensive.

Traditionally, the following procedure was used in engineering design. An engineer used mathematical and scientific theories to develop the design of a part or system. The design was then transferred to a drawing that showed the dimensions. A machinist used the information provided by the drawing to create a prototype of the part. This prototype was tested in conditions similar to those the actual part would be subjected to during its service life. Invariably, problems with the design were discovered during testing (cracks developed in certain regions, parts did not mesh well together, the function requirements of the system were not met, and so on). This information was used by the engineer to modify the design. New drawings were created and prototypes made again. When the new prototypes were tested a new set of problems might be discovered. This process was repeated as necessary until the system was complete and functioned according to the design specifications.

The completed design was given to a manufacturing engineer to determine how the actual part or system would be mass produced. Many times, the new design warranted a retooling or other significant change in the machines used to manufacture the new part. Thus, the initial costs associated with a new design would be significant. At this point, however, it was too late to go back and redesign the entire system to consider the manufacturing costs. This old design paradigm made new designs very expensive and led to a new design paradigm called concurrent engineering. Computers played a pivotal role in the development of this new paradigm, resulting in significantly lower design costs.

1.2 Computers in Design

Computers have been used in engineering design for several decades. In the early years, computers were used primarily for their "number-crunching" capabilities. In other words, computers performed the many tedious calculations involved in engineering design. Over the years, the role of computers in the design process has changed significantly. Graphical computer workstations evolved and with them, the ability of engineers to "see" their design before building it. Engineers can also do much of the testing of the design "on-screen," eliminating the need for numerous prototypes. Numerical methods such as finite-element analysis, modal analysis, and thermal analysis have enabled engineers to design systems in a fraction of the time when compared with traditional design methods. Many modern design software packages are also easily incorporated into the manufacturing process directly, enabling the designer to establish cutting-tool paths on-screen for the efficient manufacture of computer-generated models. Other manufacturing capabilities such as rapid prototyping, where physical prototypes can be created within a matter of hours rather than days, have been a direct result of computer-aided design capabilities.

Today, Computer Aided Design (CAD) is an efficient design method. The basis for this design is the construction of a graphical, three-dimensional model on the computer. This model can be "tested" by any of the available numerical methods. Design modifications can be accomplished on-screen and the modified 3-D models tested again. When the engineer is satisfied that the design will meet or exceed all of the design criteria, a 2-D drawing is created using the 3-D part as the basis. From this drawing, a physical prototype is created and tested by traditional means to ensure compliance with the design criteria. From here, the design is handed over to the manufacturing engineering division for mass production. Alternatively, if a Computer Aided Manufacturing (CAM) system is available, the part or parts can be produced directly from the 3-D computer models.

1.3 Concurrent Engineering

As CAD evolved over the years, a new paradigm developed in design known as concurrent engineering. Concurrent engineering is sometimes referred to as "design for manufacturability." With concurrent engineering, designers, analysts, and manufacturing engineers work together from the initial stages of the design process. In this way, each applies his or her expertise to the problem from the start. Thus, early in the design process, the manufacturing engineer could say, "If you made this minor modification to the part geometry, we would save

$100,000 in retooling costs." This design change could be easily implemented during the initial phases of the design, whereas without concurrent engineering, this change (and the related cost savings) may be impossible to incorporate into the design at a later stage.

Modern computer workstations have enabled concurrent engineering to become a reality in many workplaces. With local area networks (LANs), wide area networks (WANs), and the Internet, data can be moved from one desktop to another almost effortlessly. Members of the concurrent engineering team who are working in different countries can share design ideas as easily as those working within the same building. Using the principles of concurrent engineering, manufacturers can save thousands and even millions of dollars. In addition, computer-aided, concurrent engineering design is more efficient, and the results are often of higher quality when compared with the designs produced in the past.

1.4 Flexibility in Design

Early CAD systems were relatively cumbersome to use, but they have vastly improved in recent years. The key advantage of modern CAD systems is their ability to produce flexible designs. Changes can be made easily to computer-generated models, and engineers can play "what if" with their designs. They now have the ability to ask themselves, "What will happen to the performance of the entire system if I make this one minor change to a single part in the system?" In this way, they can optimize the design to obtain the best possible balance between cost and functionality.

One advantage of CAD software that has led to highly flexible design is the concept of feature-based modeling. With feature-based modeling, an object is considered to be made up of features. For example, the object shown in Figure 1–1 is made of two features—an overall block and a cylindrical hole removed from it.

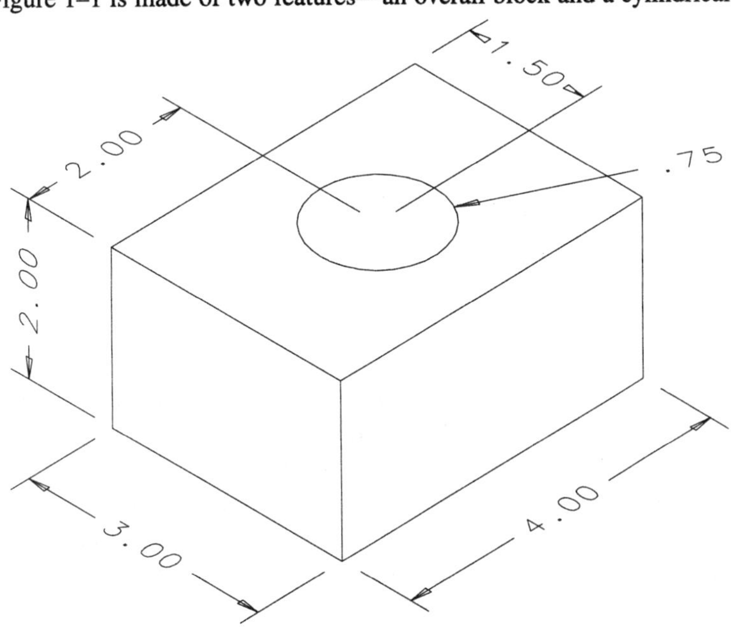

Figure 1–1 Object Composed of Block and Hole Features

Furthermore, the location of one feature (the hole) is defined relative to the edges of the other feature (the block). Thus, instead of a being considered a single object with the dimensions as shown, this basic design concept can also be used to describe the object shown in Figure 1-2. Note that for the object shown in Figure 1-2, the dimensions of some of the features have changed, but the basic design concept has not changed. It is still a block with a cylindrical hole in it, and the location of the hole is defined relative to the edges of the block. There are an infinite number of possible objects that can be created from this single design concept. In this way, it is clear that the design concept (also called the design intent) is of paramount importance, and the actual dimensions of the object are merely details that can be accurately defined at a later stage in the design process. Various software packages have different approaches to feature-based modeling. Some use a concept called "parametric modeling," and others are based on "variational geometry." Each of these approaches is slightly different, but they both lead to highly flexible design. The difference between the two approaches will be described later in this text. Whichever approach is used, the end result is a flexible design defined by its features.

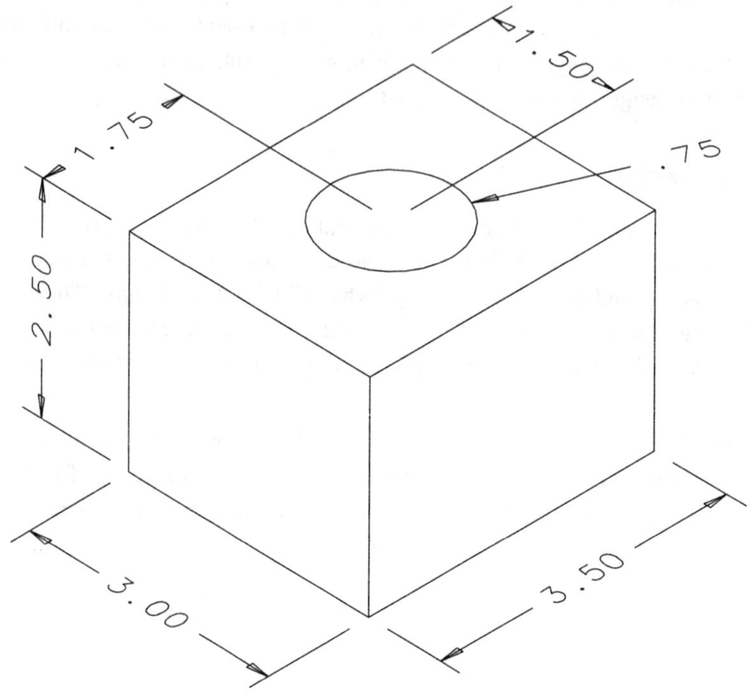

Figure 1–2 Object Composed of Block and Hole Features

1.5 Text Organization

This text is organized such that each chapter (except for Chapter 2) contains a generic description of the topic under discussion. This description is not specific to Pro/ENGINEER software, but is meant to be an introduction to the concepts that are common among most 3-D solid modeling packages. This introductory description is followed by specifics that relate to Pro/ENGINEER software only. For example, in Chapter 5, there is a general description of how 3-D objects can be displayed followed by specific instructions related to the use of Pro/ENGINEER in accomplishing these tasks. Interspersed within the description of the use of Pro/ENGINEER software are "Guided Tours." The Guided Tours are your opportunity to use the software to achieve the stated objectives. If you complete the Guided Tours, you will have a good understanding of the workings of the software. You should try to work as many of the Guided Tours as you can in order to facilitate your learning the software.

Chapter 2 consists of basic information regarding user interaction and management of files and parts within Pro/ENGINEER. While it may seem premature to discuss file and part management before you have learned about actually creating parts, this chapter will serve as reference material for you to consult as you learn about managing your work. The remaining chapters are organized around certain tasks that you will perform in creating 3-D models. After introducing the new topics and the Guided Tours, there is a strategy section (except for Chapter 2) followed by a troubleshooting/FAQs (frequently asked questions) section for each chapter. At the end of each chapter, there is a summary of key concepts followed by exercises for you to test your mastery of the topics from that chapter.

This is not meant to be a comprehensive text on the use of Pro/ENGINEER software. Rather, it is meant to serve as an introduction to some of the basic procedures in three-dimensional modeling with Pro/ENGINEER. For this reason, many of the advanced features of the software are not covered in this text. If you are interested in learning more, you should consult the on-line help or other training materials available directly from Parametric Technologies.

This is also not a "cookbook" type of manual. The only way you will really learn this software is if you **sit down and experiment with the various available commands**. Hopefully this text will be a useful guide for you and will facilitate your learning. Good luck.

CHAPTER 2

User Interface and File Management in Pro/ENGINEER Software

Pro/ENGINEER is a fully integrated, multifaceted engineering tool that is developed and distributed by Parametric Technology Corporation. It is a comprehensive suite for design, analysis, and manufacturing. Pro/ENGINEER software enables you to create parts, assemble these parts into systems, test the structural integrity of the individual parts, create multiview drawings from the parts or assemblies, and develop cutting-tool paths for machining purposes or bends and welds for sheet-metal applications. In addition, recent releases of the software have included extensive tools for the management of the information contained in the engineering design. One of the key advantages that Pro/ENGINEER has over similar software available on the market is that it is a fully integrated package. The geometry needs to be created only one time, and it can be readily transported to the analysis and manufacturing tools within the software. There is no need to translate the geometry so that the output file structure is compatible with the input requirements of an external software package. A significant amount of time can be saved in performing the engineering analysis using this integrated tool.

Pro/ENGINEER software is logically divided into several sections by integrated file types. The major types include sketch, part, manufacture, and drawing as well as format, report, diagram, layout, and markup. The files are fully integrated, and changes in one file automatically update the others. Pro/ENGINEER's updates are bidirectional. This means that, as a part is created and pulled into a working drawing, any changes made to the drawing will also update the base-part file. In order to utilize the features of Pro/ENGINEER, you must first be able to create part geometry. Likewise, in order to perform simulations or develop manufacturing tool paths, you must first create a part. The ability to create a part is the foundation upon which all the other engineering analysis activities are built. Therefore, it is critical that you understand the fundamentals of part creation before you can work further with a software package such as Pro/ENGINEER. This text focuses on the basics of part creation. It is not meant to be a comprehensive text on the use of Pro/ENGINEER software as a design, analysis, and manufacturing tool.

The purpose of this chapter is to familiarize the reader with the basic working environment of Pro/ENGINEER, to allow the reader to become aware of specific commands in Pro/ENGINEER, and to serve as a quick reference guide. It is not the goal of this chapter to cover every aspect of Pro/ENGINEER's working environment.

2.1 Pro/ENGINEER's Working Environment

You typically start Pro/ENGINEER by typing *pro* or *proe* at the prompt line. However, the start-up procedure used in your lab setting might be slightly different. Ask your lab assistant, instructor, or system administrator for help in how to get started with the software. Once you have successfully launched Pro/ENGINEER, a base window should appear similar to the one shown in Figure 2–1.

The working environment of Pro/ENGINEER is similar to many windows-based products. The base window is divided into several convenient regions. The largest region is the main graphic area, where the majority of time will be spent creating your parts. Those familiar with typical windows applications will recognize the standard pull-down menu tool bar and the shortcut tool bar. The bottom region of the page contains the message window and the command line (with Pro/ENGINEER 2000 the command line is found directly under the shortcut tool bar). Each of these regions is expanded upon below.

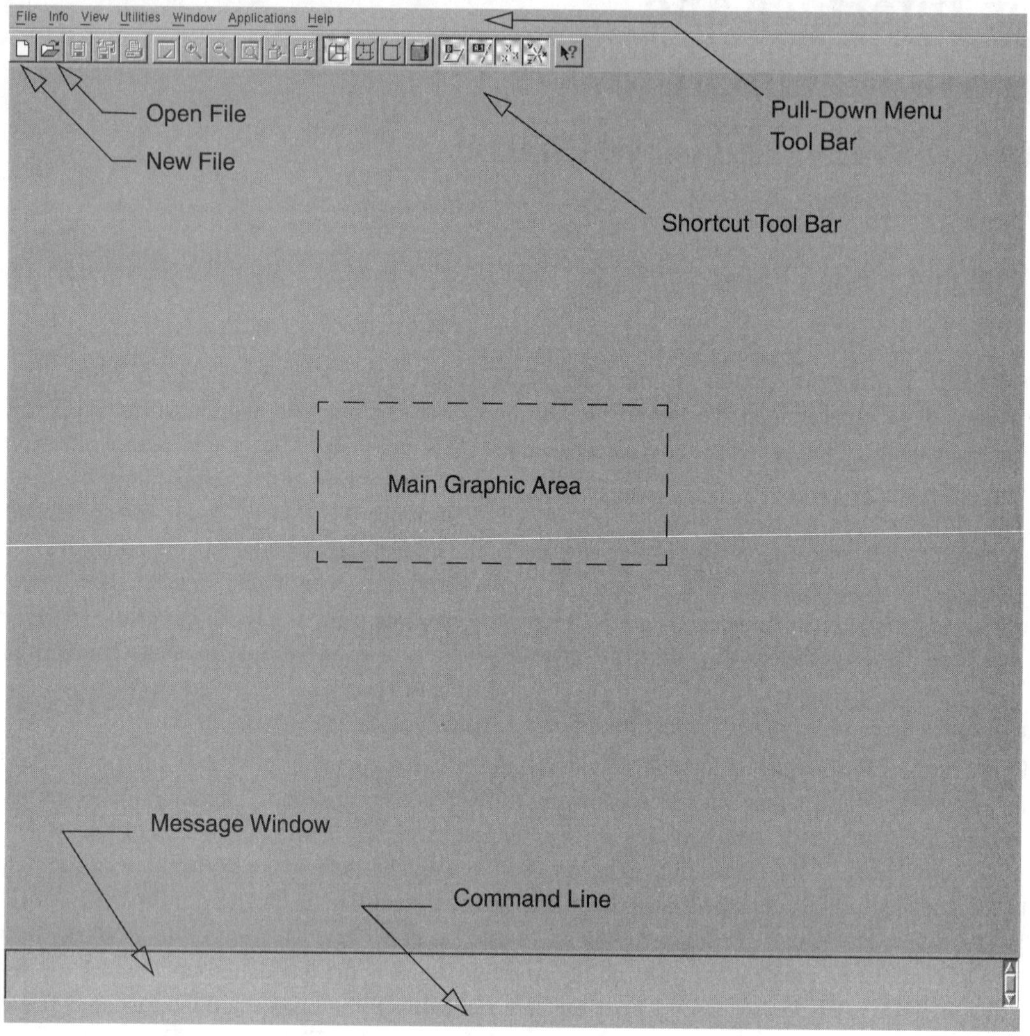

Figure 2–1 Pro/ENGINEER's Base Working Window

A word of caution, messages periodically appear to confirm that a selection has been made, to notify you of errors or potential problems, or to request information. At times, your mouse may seem to stop working or appear to be "locked up." If this happens, check the message window to see if Pro/ENGINEER is requesting an input.

2.1.1 The Pull Down Menus

The software engineers at Parametric Technology Corporation put great effort into designing a software package that is powerful, easy to use, and quick to learn. The user may find that many of the pull-down menu operations are similar to other common computer applications. While several of the operations are common, however, others may be new and, if used without a complete understanding, can ruin your day with an accidental delete. To avoid this catastrophe, File and Window, two of the main pull-down menus, are highlighted in this section. The remaining pull-down menus will be explained as the need arises.

File: To activate the first pull-down menu, move the curser up to *File* and press the left mouse button once. Several commands found on this menu are as follows:

New Use this command to build a new part, start a new sketch, develop a new drawing, etc.... After selecting this command, a dialog box will appear near the center of the graphic area on Pro/ENGINEER's base page.

Open Use this command to open a previously created file without exiting Pro/ENGINEER. When *Open* is selected, a new working or base page will appear with the selected part, sketch, or drawing on it. The model or part you were working on before selecting *Open* is located on a separate base window beneath the present window.

Working Directory This command allows you to select the directory in which you are going to store and retrieve parts. This is a handy command when you are working on multiple projects.

Erase Use this command to "clear the memory" of your working session. The use of this command will not remove any files saved in your directories. This is a somewhat advanced command.

Delete Use this command to remove an old version of your part. Be careful with this command. If you select *All versions* at the prompt, no copies will remain in your directory. It is best to select *Old versions*, which leaves a copy of the most recent part in your directory.

Save Use this command to save your work. Pro/ENGINEER will NOT prompt you regarding unsaved work.

Save As Similar to the Save command, this command saves your work under a different file name. Two files will exist after this command is used, the old file and the new file. Pro/ENGINEER assumes you wish to continue working with the old file.

Backup Use this command to create a backup file. This is useful when working on long-term projects in which several revisions of your part will take place.

Rename Use this command to rename your file.

Import This command is used to import work from a file that is not in standard Pro/ENGINEER format. Thus, you can import data created in other programs to Pro/ENGINEER. The available external file formats for import include: IGES, SET, VDA, and CADAM, as well as several others.

Export Similar to the Import command, this command is used to write data for use in other software. Pro/ENGINEER is able to export data in common format, images in tiff, jpeg, and eps. The software also has a direct VRMC output option.

Print Use this command to obtain a hard copy of the image on the screen. Since there are a wide variety of methods by which printing is handled on different networks, you must discuss the printing setup with your lab consultant or system administrator.

Exit With this command, you exit Pro/ENGINEER. A prompt will ask if you really want to exit, but all unsaved work will be lost. Save before selecting this command!

Window: One basic menu that will become more important as we continue is the *Window* menu. This menu allows us to activate, generate, or close windows. It also allows us to toggle between several working parts or models.

Activate Use this command to activate the frontmost window. This is often used when working with multiple parts or models.

New This command is used to generate another view of the current part. Any modifications to the part will appear in all related windows.

Close This feature is similar to the close feature found under the file menu of many common applications. It closes the present window. WARNING: Pro/ENGINEER will not prompt you to save any unsaved modifications, so SAVE BEFORE YOU CLOSE!

Model Tree Use this command to toggle the model tree window on and off. This command is helpful for retrieving the model tree if it gets hidden beneath other windows.

Open System This feature will not be discussed in this text.

Index This lists all the open windows and indicates the active window with a small bullet.

A basic understanding of the File and Window menus will enable you to start building parts and developing models. As your skills improve, your ability to use the commands discussed will increase your efficiency.

While only the File and Window pull down menus were discussed here, the other menus provide valuable information as well. As we progress through this book, more commands in the pull-down menus will be discussed. Many of these features can be found in the shortcut tool bar.

2.1.2 Shortcut Tool Bar

The shortcut tool bar is divided into the following four segments: file operations, screen view operations, image-style operations, and datum-plane operations. Once you become comfortable with the icon-based shortcuts, less time will be spent using the pull-down menus.

File Operations: The shortcut group for the file operations are shown in Figure 2–2. From this group, files can be created, opened, saved, saved with new names, and printed. The use of this icon group will increase efficiency and save time.The description for each operation is the same for the shortcut buttons as for the pull-down menus.

View Operations: The shortcut group for the view operations include repaint, zoom controls, refit, orient, and view list. These commands will be used in the development of most parts or models you build.

- *Repaint* clears the screen and redraws the object. This helps to remove fragments of line that were not completely redrawn following a previous command.
- *Zoom In* allows you to draw a box around a section you wish to examine more closely in the main graphic area. It is good practice to select *Repaint* following a *Zoom In*, as this ensures that the *Zoom In* option is no longer active.
- *Zoom Out* allows you to slowly increase your viewing distance. This feature reduces your view by a factor of two each time it is activated.
- *Refit* will center your part in the main graphic area and zoom in or out such that the entire part plus a small margin surrounding the part will fit in the main graphic window.
- *Orient* is used to align your part orthographically.
- *View List* is used to select from a list of previously stored views. The default view is automatically included in this list.

File Operations

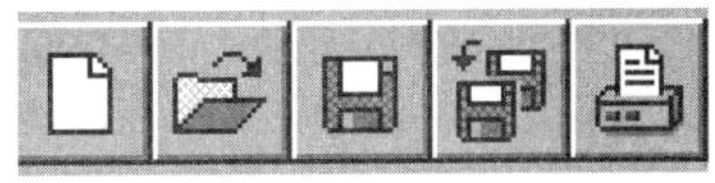

New I Open I Save I Save As I Print

View Operations

Repaint I Zoom In I Zoom Out I Refit I Orient I View List

Image Style Operations

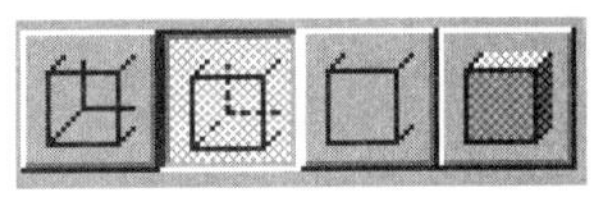

Wireframe I Hidden Line I No Hidden I Shaded

Datum Plane Operations

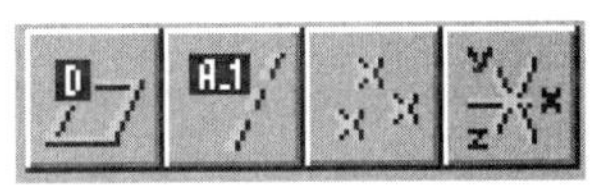

Datum: Planes I Axis I Points I Coordinate Systems

Figure 2–2 Pro/ENGINEER's Shortcut Tool Bar

Image Display Styles: The shortcut buttons, which control the way the image of the part is viewed on the screen, are the Image Style operation buttons. These allow the designer to view the part in wire frame (Figure 2–3A), with hidden lines (Figure 2–3B) without hidden lines (Figure 2–3C), or as a shaded image (Figure 2–3D). It is generally good practice to construct your object in hidden line mode. This allows you to see hidden features and select hidden surfaces. Shaded images mode allows you to get a "realistic" view of the part you just created. This is a powerful mode, but it will slow down your computer, thus, it is good practice to avoid printing the shaded images.

A B C D

Figure 2–3 Different Image Display Styles Available in Pro/ENGINEER

Datum Operations: Datums, also know as object aides, play an important role in the design and development of any model or part in Pro/ENGINEER. Not only are datums used as references in dimensioning, but they are also used to align, cut, rotate, and revolve features. When building your model, you should keep all datum plane features on. This will help you reorient your part and ensure you always have references available. Figure 2–4 shows the different object aides available by toggling the datum operation shortcut buttons.

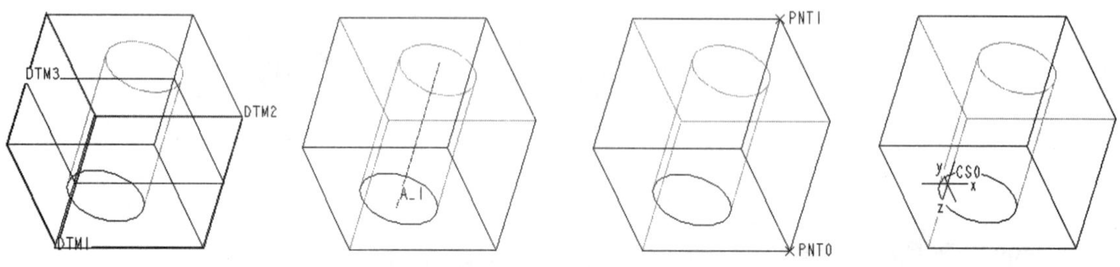

Datum Planes — Datum Axis — Datum Points — Coordinate System

Figure 2–4 Datum Operations, Examples of Object Aides for Pro/ENGINEER

The commands, shortcuts, and menus discussed previously serve as building blocks from which we will develop and expand our skills. Keep this section handy as we will refer back to it from time to time. Now, we will set up the working environment in Pro/ENGINEER.

Guided Tour 2.1

- Start up Pro/ENGINEER and select the new file icon from the shortcut tool bar.
- In the dialog box that pops open, verify that *Part* is selected and that *prt0001* is entered as the name. Once verified, press *OK.*
- Two new dialog boxes should now be open, Model Tree and Menu Manager. In the Menu Manager, select: *Feature—Create—Solid—Protrusion—Extrude—Solid—Done.*
- From the Sketch Menu Manager, select: *Circle—Geometry—Center/Point.*
- Move to the main graphic area. With the left mouse button, click once, then move the mouse and click again. You have just sketched a circle.
- From the Main Menu Manager, select *Dimension*, then left click on the circumference of the circle (the circle should now appear red). Move the mouse about an inch away, and with the middle mouse button, click in a clear area of the screen. You should now see a radius dimension label *sd0.*
- From the Main Menu Manager, select *Regenerate*. The dimension, *sd0*, should now have a numerical value, and the Menu Manager should now be in Modify mode. Left click on the dimension value, and enter a new value of *10.*
- Select *Regenerate* from the Main Menu Manager. The circle should now have changed size and the screen self-adjusted to fit this new size.
- From the bottom of Main Menu Manager, select *Done* and enter a depth of 20. Select *Preview* from the Feature Elements Window, then select *OK.*
- From the shortcut tool bar, select *Save—View List—Default—Print.*
- Congratulations! You have just created your first Pro/ENGINEER model.

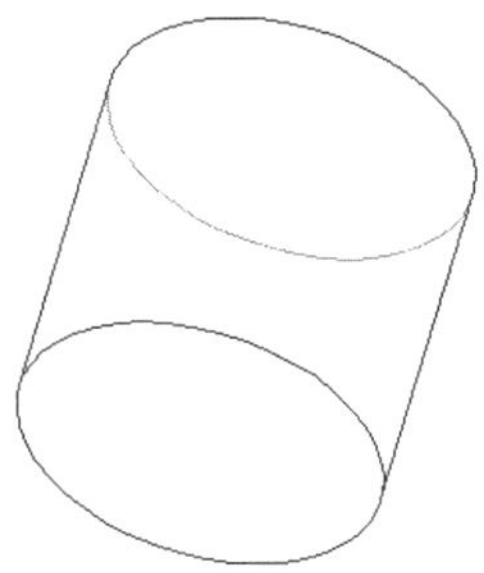

2.2 The Modes in Pro/ENGINEER

Using Guided Tour 2.1, you should have successfully created your first part using Pro/ENGINEER. You did this in the Part mode selected from the New File dialog box, Figure 2–5. Pro/ENGINEER refers to the different working environments as *modes*. For this text, we will focus most of our attention on the following four most commonly-used modes:

Part The Part mode is used to create and modify parts and models. This mode incorporates the use of the Pro/ENGINEER Sketch mode.

Sketch The Sketch mode is used to create sketches and cross sections for extrusion. It is also used as an engineering "scratch pad."

Drawing The Drawing mode is used to create finished engineering drawings of previously created parts and assemblies. The finished details such as section views and dimensions are added in this mode. (See Chapter 8.)

Format The Format mode is used to create custom title blocks and drawing borders. From the Drawing mode, you will be able to retrieve previously created formats. (Note: Text added to a title block in Format mode cannot be edited in Drawing mode. Delay adding text that needs to be edited until you are in Drawing mode.)

Working Modes of Pro/ENGINEER

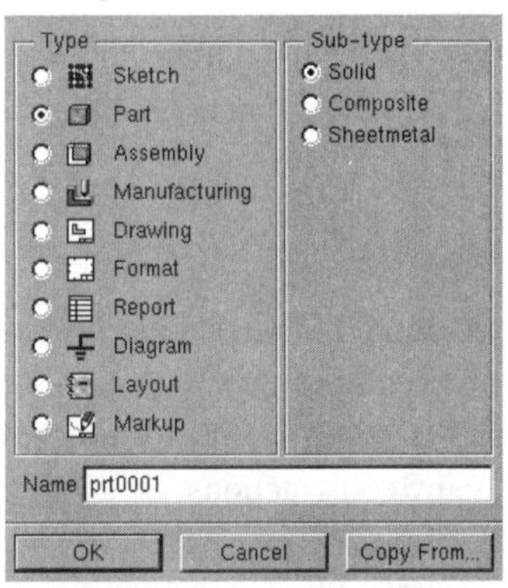

Figure 2–5 New File Dialog Window

An understanding of the features and operations in the Part mode will enable you to create assemblies, drawings, and formats in their respective modes.

2.3 Setting Up a Working Environment in Part Mode

Once you have successfully launched Pro/ENGINEER in Part mode, a menu and a dialog box appear. These are the Model Tree and Menu Manager, Figure 2–6. The Model Tree serves as an outline for your part. Each of the features added to your part shows up in this menu. The Model Tree can aid in the selection of parts to modify or delete. It is often easier to select a feature from a list of items than from a picture of the part itself. The Menu Manager houses all the commands needed to create, modify, and delete features. Becoming familiar with the Menu Manager and how it operates will improve your ability to create solid models and design parts using Pro/ENGINEER.

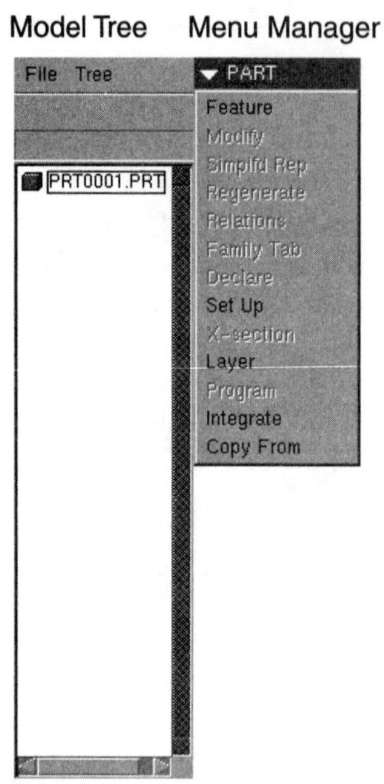

Figure 2–6 Model Tree and Menu Manager

2.3.1 Unit System

The selection of *Set Up* gives you the ability to define many options. Most of these options are beyond the scope of this text. Since you will use the *Units* option regularly, we will discuss its use. When the *Units* option is selected, the Menu Manager expands to include *Principle System*, *Custom Units*, and *Scale Model*.

Principle System Use this command to select or define the units of length, mass, time, and temperature to be used throughout the development of your model. Units can be specified after a part has been created, however, it is better to define units before beginning any constructions.

Custom Units Use this command to define your own unit system.

Scale Use this command to change the building scale of your part or model. While this command is available, it is good practice to create everything in full scale (1:1).

2.3.2 Datum Planes

After defining your unit system, your next procedure is to create reference datum planes. To define the reference datums required to create complex parts, you use the following sequence of commands:

Feature—Create—Datum—Plane—Default

In the Model Tree, you should now see three features labeled *DTM1*, *DTM2*, and *DMT3*. Three yellow datum planes should also appear on your screen. If the datum planes are not visible, verify that the Datum Plane button is activated (look at the shortcut buttons). These datum planes are now a feature of your part. When you start to create other features, Pro/ENGINEER is going to ask you to locate the section or feature with respect to the part. To do this, you will need to give location dimensions from the datum planes to the new section or feature.

It is also good practice to create a coordinate system using the following sequence:

Feature—Create—Datum—Coordinate System—Default—Done

This sequence of commands sets up a right-handed x-y-z coordinate system. In the Model Tree window, *CSO* should now appear under the datum planes. If it does not appear, you were not successful in creating the coordinate system. To correct this, simply run through the command sequence again.

Guided Tour 2.2

- Start Pro/ENGINEER, select *New* from the File menu. Select *Part* mode from the New File dialog box. You may wish to change the file name, but for this example, we will stick with the default name of *prt0001*.
- Before we build a part, the working space must be defined. While not absolutely necessary at this time, it is good practice to start using datum planes. Datum planes act as references for building and constraining your part or model. Once created, datum planes will always be a "feature" of your model. To create datum planes from the Menu Manager, use the following sequence of commands:

  Feature—Create—Datum—Plane—Default

- Once you click on *Default*, three yellow planes should appear in the main graphics window. The Model Tree should also show three items labeled *DTM1*, *DTM2*, and *DTM3*. The datum planes are now *The Part*. Let's add something to these datum planes.
- To create a simple extrusion, we are going to follow a command path similar to the one we used to create the datum planes. The command sequence is
  Feature—Create—Solid—Protrusion—Extrude—Solid—Done
- After selecting *Done*, you should see a new window similar to the following figure.

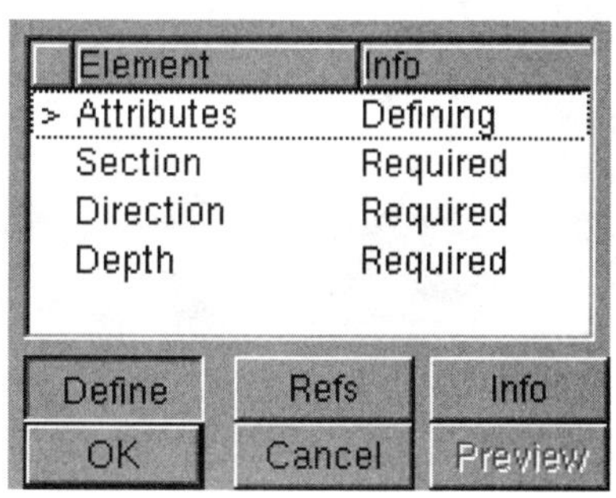

- This window is called the Feature Elements Window. Each element must be defined in order to have the feature we wish to create become incorporated into the part. Pro/ENGINEER will automatically move to the next element once the current element has been defined. Below the Feature Elements Window you will see the Menu Manager asking for the attributes of this feature. For this example, select

  One Side—Done

- Notice that the arrow in the Feature Elements Window has moved to Section and the Attributes are now fully defined. The Menu Manager has also changed as Pro/ENGINEER requires that we define the section we are going to extrude in order to build our part. To completely define the section, we need to select a sketching plane, sketch the shape to be extruded, and fully constrain that section.

- To select the sketching plane, use the following command sequence:

  Setup New—Plane—Query Select

- Click the left mouse button on the text of datum plane 3 and click DTM3. A cyan colored box will appear around the datum plane. The message window will ask you to confirm the selection, and the Menu Manager will ask you to accept. Select *accept*. It is good practice to use Query Select when selecting any part, edge, feature, etc.

- At this point, you should notice an arrow sticking out of datum plane 3. The Menu Manager is prompting you for a *Flip* or an *OK*. Select *Flip* a few times to see what happens. When you are happy with the direction select *OK*.

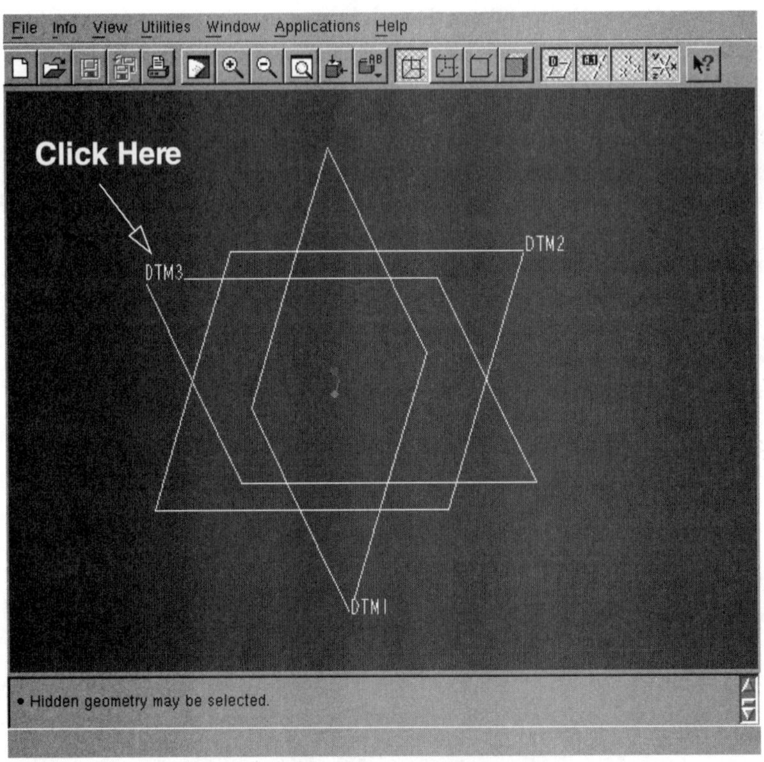

- Once the direction has been selected, a second reference plane needs to be defined. In the Menu Manager, select

  Top—Plane—Query Select—(then click on the text *DTM2*)

- Once the proper plane is highlighted, select *accept*. Pro/ENGINEER automatically reorients the datum planes and switches into Sketch mode. Close the information window that pops up in the middle of the screen.

- Look at the Menu Manager. Verify that the Intent Manager is not check marked. You should see Sketch—Mouse Sketch highlighted. We are going to draw a circle, so select

  Circle—Geometry—Center/point

- Click in the upper right quadrant, and sketch a small circle similar to what you did in Guided Tour 2.1. Recall that before the section can be extruded, it must first be regenerated. In the Menu Manager select *Regenerate*.

- The message "Locate section with respect to part" should show up in the message window. Recall that Pro/ENGINEER thinks of the datum planes as the part, and we need to locate the section with respect to the datum plane. One way of doing this is to dimension the location. There are two basic types of dimensioning required by Pro/ENGINEER, one is size and the second is location.

Creating the sketch

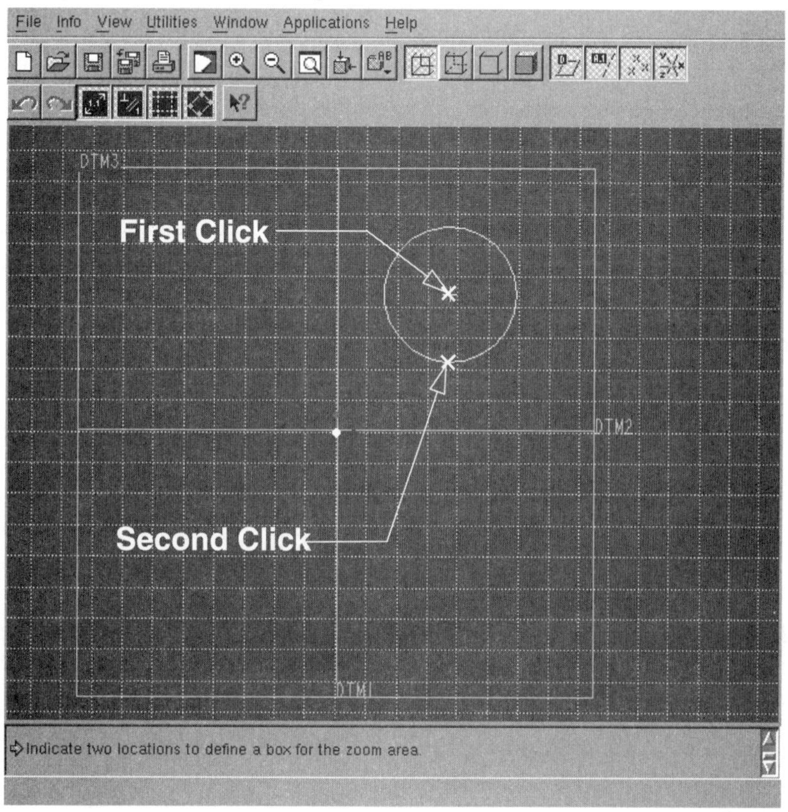

- Dimension the location of the circle by using the left mouse button, and clicking the center point once (the center point should now be red). Then, again with the left mouse button, click the vertical axis/datum plane once (it should be highlighted). Now, with the middle mouse button, click about halfway between the first and second clicks. Your window should now look like the next figure. The dimension "sd0" is now helping to locate your section to the part. Do the same procedure for the horizontal axis. Remember the sequence "left click, left click, middle click." Once you have located the section with respect to the part, select *Regenerate* from the Menu Manager.

Dimension the feature

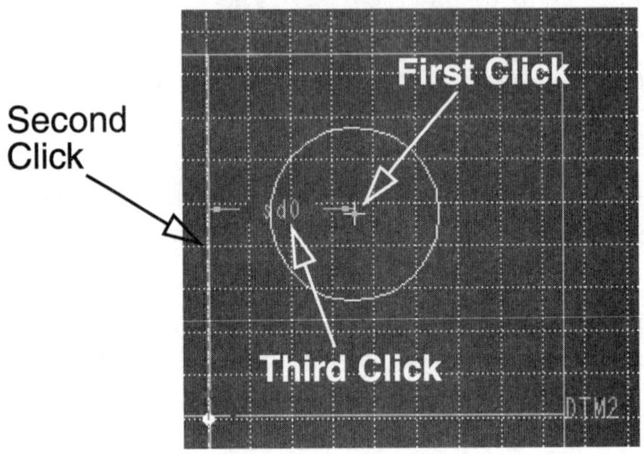

- A new error message should appear in the message window, "*Underdimensioned section. Please align to part or dimension.*" The circle is now red indicating that it is underdimensioned. To fix this problem, verify that you are in *Dimension* mode and left click the circumference of the circle, move about an inch away and middle click in a blank area ("left click, middle click"). You should now see an "sd2" dimension. Now select *Regenerate* from the Menu Manager.

- Modify these values to something more reasonable. Verify that you are in *Modify* mode on the Menu Manager, and select the dimension for the radius of the circle with a lift click. Enter a new value of *10*. Then select the two location dimensions and change them both to *20*. All three dimensions should now be a white color. Hit *Regenerate,* and see what happens.

- The screen should have changed appearance to zoom in or out, and we should see a note in the message window, "*Section successfully regenerated.*" We are finished defining this section, so select *Done* from the Menu Manager.

- Now, you must define the depth. Select

  Blind—Done—(enter a value of *10*)

- To ensure we are working with the correct shape, go up to the Feature Elements Window and select *Preview*, then from the pull-down menu, *View*, select *Default*. Your screen should look like the following figure:

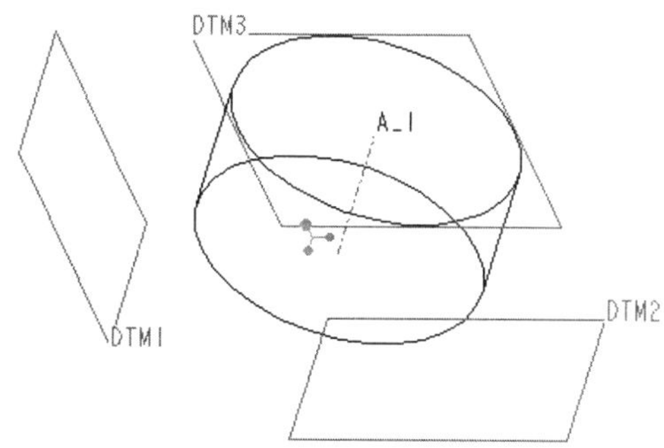

2.4 Preferences in Pro/ENGINEER

Pro/ENGINEER has the flexibility to suit many of your personal preferences. The two areas we will focus on in this section are the Environment and the Color.

2.4.1 System Environment

In the previous section, we discussed the setup of the working environment. This included selecting a unit system and generating reference planes. In this section, we focus on the modifications that you can make to the system environment. A selection of *Environment* from the *Utility* pull-down menu reveals many of the simple modifications that can easily be made to suit your preferences. Many of these options are also found on the shortcut tool bar.

The Environment Window is divided into three sections. The top section is where modifications to the display are made. This area includes all the datum operation options, and other more advanced operations. One option that is frequently turned off is the *Spin Center.* The *Spin Center* acts as a reference when the orientation of the object is changed. This feature becomes less important if a default coordinate system was defined during setup.

If you are easily annoyed by the miscellaneous beeps that sound as you make selections, you can turn the sound off in the *Default Action* section of the Environment Window. By selecting *Ring Message Bell*, you will no longer be audibly prompted for each message. This also means that you will have to pay closer attention to the Message Window at the bottom of the screen.

The lower section of the Environment Window serves as the location for the *Display Style, Default Orientation*, and *Tangent Edges*. The *Display Style* was discussed in Section 2.1.2 and shown in Figure 2–3. The display can be changed among wireframe, hidden line, no hidden lines, and shaded images. These options are also available in the shortcut buttons.

The *Default Orientation* is set to *Trimetric*. In this view, Pro/ENGINEER keeps the x-axis horizontal and rotates the y- and z-axes so all three dimensions will appear on the screen. Many users change the default from *Trimetric* to *Isometric*. *Isometric* aligns the y-axis in the vertical direction and rotates the part such that the coordinate axes appear 120° apart. The user may also define his or her own orientation. (Note: While the isometric view is often considered an industry standard, this view does not work well in the part/model development phase due to the overlap in the datum planes area.)

The conventional practice is not to show the tangent lines associated with large-diameter arcs. However, when creating objects in Pro/ENGINEER, it is sometimes useful to be able to see where the point of tangency is located. For this reason, the default option of *Tangent Edges* is set to *Solid.* For finished drawings this should be set to *No Display.*

These options are left up to you to explore so that you can find the preferences which best fit your style. For the Guided Tours throughout this book, the *Spin Center* will be turned off, the *Default View* will be set to *Trimetric*, and the *Tangent Edges* will be set to *Solid.*

2.4.2 Color Preferences

Color preference varies widely from person to person, so Pro/ENGINEER has capacity to customize the colors found in the working environment. Under the *Utilities* pull-down menu, you will find *Color* with the options of *System* or *Entity.*

System Colors This option enables the user to redefine the colors for the background, lines of the geometry, hidden lines, lettering, etc.

Entity Colors This option enables the user to change the color of the datum planes, axes, points, and coordinate systems. This command does not change the shaded color of the object. (Changing the shaded color of the object is covered in Chapter 5.)

Guided Tour 2.3

- Start a new part, *prt0002.*
- Change the units to millimeters, Newtons, and seconds.
- Add a set of default datum planes.
- Change the background color to red, then change it back to blue.
- Change the environment preferences so the default orientation is *Isometric*

2.5 Pro/ENGINEER's Intent Manager

Pro/ENGINEER has an effective tool to help you become more efficient with your model development. This feature is called Intent Manager. Intent Manager is available when you are in the Sketch mode and automatically dimensions and constrains your sketch. We will discuss dimensioning and constraining further in the next chapter, but the reader is encouraged to experiment by creating a few sketches with Intent Manager turned on or off, with and without first creating datum planes.

2.6 Obtaining On-Line Help in Pro/ENGINEER

Like most modern software, Pro/ENGINEER has on-line help available to you as you are working with the software. In Pro/ENGINEER the on-line help is accessible simply by selecting *Pro/E Help* from the Help pull-down menus. This option should automatically load a web browser (see your local system administrator for details). The on-line help relieves you of the need to interrupt your working session to consult a manual when you get stuck. As you are learning how the software works, the on-line help will be very useful.

2.7 Chapter Wrap-Up

After reading this chapter you should

- be familiar with the working environment of Pro/ENGINEER, including the pull-down menus and the shortcut tool bar,
- understand that Pro/ENGINEER has different modes for the creation of models. These modes include Part, Drawing, Assembly, and Format,
- understand how to set the units and create datum planes, and
- know how to modify the system environment to match your preferences.

2.8 Chapter Summary

In this chapter, you learned how to start Pro/ENGINEER. You also learned about the user interface in Pro/ENGINEER and some basic methods for user interaction within the software. Access to the on-line help in Pro/ENGINEER was also discussed.

Exercises

1. Start up Pro/ENGINEER and select New File. How many modes are available? Write a list of the available modes in the order in which they appear. Which modes have sub-types?
2. Load the part from Guided Tour 2.1. In the main graphic window, with the control key held down, click with the left mouse button and drag the pointer across the screen. What happens? Do the same with the middle mouse button and the right mouse button, what happens?
3. In the sketching phase of Guided Tour 2.1, four new shortcut buttons appeared. What are their functions? (You might want to restart Guided Tour 2.1 and slowly slide the pointer across the icons until the description message appears.)
4. Select *Pro/E Help* in the Help Menu. What happens?
5. Restart Guided Tour 2.2, only this time, turn on Intent Manager.
6. Create a cube, 10 x 10 x 10, following the directions given in the previous Guided Tours.

CHAPTER 3

Two-Dimensional Constraints

The earliest versions of solid modeling software provided little opportunity for you to change your existing geometry easily. Designers were forced to start over in order to incorporate even the smallest change in the details of part creation. Fortunately, software today is very flexible, and you are able to effect changes in your design as necessary. The source of this flexibility stems from the use of constraints within solid modeling software. The remainder of this chapter will focus on constraints—what they are, why they are useful, and the types of constraints typically found in solid modeling software. You will also learn how to apply and modify constraints within Pro/ENGINEER to assist you in creating parts that are the correct size and shape.

3.1 What Are Constraints?

You include constraints on your 2-D sketch to control how the geometry will behave when modifications are made to the sketch or to the model that you create. In this way, your design intent will be preserved even as size modifications are made to the part later on. When you lived at home with your parents, they applied constraints to control your behavior. Some of their constraints might have included your curfew, your bedtime, the amount of TV you were allowed to watch, what rating of movie you were allowed to see, and so on. As you grew, the number value associated with the constraint (e.g., your bedtime) changed accordingly, however, your behavior was still constrained by your parents' specifications. Constraints placed on sketches act in much the same way. If you constrain two lines to be parallel, then no matter what happens, if you or someone else changes a dimension on the geometry, those two lines will remain parallel to one another. If you constrain a line on the geometry to be vertical, then that line will remain vertical through all modifications made to thc part. If you constrain the distance between two lines, you can go back later and change that distance or you can leave it as it is.

There are two major types of constraint-based modeling systems available on the market today: **variational geometry** and **parametric modeling**. With variational geometry, constraints are applied to the 2-D shape at the **degrees of freedom**. A degree of freedom exists if there is currently no constraint applied to control the geometry. For example, if you have two lines on your sketch that are completely unconstrained with respect to one another, then a degree of freedom exists. If you then constrain the geometry so that the lines must always remain parallel to one another, then that degree of freedom is removed because now the lines are constrained relative to one another. Variational geometry modelers consider the total number of degrees of freedom that your sketch contains, and as you add constraints, degrees of freedom are removed from consideration. Variational solvers work on systems of equations. Therefore, if there are initially 20 degrees of freedom for your sketch, you will need 20 constraints in order for your geometry to be fully constrained. The resulting system of 20 equations (each time you apply a constraint you are creating an "equation" for the system) must be able to be solved simultaneously, or your geometry is not valid. For example, if you had an existing constraint that forced two lines to be parallel to one another, and you then attempted to apply constraints such that one line was vertical and the other made an angle of 80 degrees with respect to the horizontal position, the variational geometry solver would not be able to solve this system of simultaneous equations. The constraint that the lines remain parallel is in direct conflict with the angular values that you attempted to specify. This would not be a valid solution, and you would be required to remove one of the three constraints from the sketch in order to proceed. Another feature of variational geometry is that you do not have to work with fully constrained geometry. Certain degrees of freedom can be left unspecified on the sketch and you can still extrude or revolve the shape to form a solid.

Parametric modeling is a subset of variational geometry. Parametric modeling allows for a flexible design in much the same way that variational modelers do. You are able to add constraints to your sketch to control how the geometry will behave as modifications are made. However, parametric modeling does not work on a system of simultaneous equations. Parametric modelers "solve" your constraints in the order that they were applied. Thus, if there were conflicting constraints specified, then the most recent constraint that was added to the geometry would govern the behavior of the 2-D shape as modifications are made. If you had two lines that you first constrained to be parallel and then constrained one to be vertical and the second to be at an angle of 80 degrees with respect to the horizontal, the result would be a vertical line and a line at 80 degrees. The first constraint you made, that the lines be parallel, would be ignored in favor of the more recent constraints that you applied. With parametric modelers, you generally need to have a fully constrained shape before you can use it to create a 3-D solid. Your resulting shape may have more than one constraint that defines the same degree of freedom, however, you cannot have any degrees of freedom that are undefined.

As you work with your solid modeling software, it is important that you understand whether you are working in a variational geometry or a parametric environment. The results that you achieve may vary greatly depending upon the type of software system available to you.

3.2 Types of Constraints

As mentioned previously, constraints enable you to preserve the original design intent even as changes are made to the model you have created. Constraints allow for a great deal of flexibility in your design, but it is important that you apply constraints properly so that the integrity of your design is maintained. It is important for you to realize that in the current level of software development, constraints can only be applied between two-dimensional geometric entities on your sketch, and you are constraining one entity relative to another one. There are two major categories of constraints, geometric and dimensional.

3.2.1 Geometric Constraints

Geometric constraints are used to define the *shape* of your 2-D geometry. Figure 3–1 illustrates the types of geometric constraints you can use to define the shape of your 2-D profile. Geometric constraints include the following:

Parallel With this constraint you define one line parallel to a second line on the sketch. If the angular orientation of one line changes, then the angular orientation of the second line will change accordingly.

Perpendicular This type of constraint is used to constrain one line so that it always meets a second line at a right angle. Again, if the angular orientation of one of the lines changes, the orientation of the second line will be modified so that the perpendicularity is maintained.

Colinear This constraint is used to establish colinearity between two separate lines or between a line and an indicated point.

Coincident Similar to the Colinear constraint, this constraint is used to indicate that two points on a sketch coincide with one another. This is especially useful in constraining two circles to be concentric.

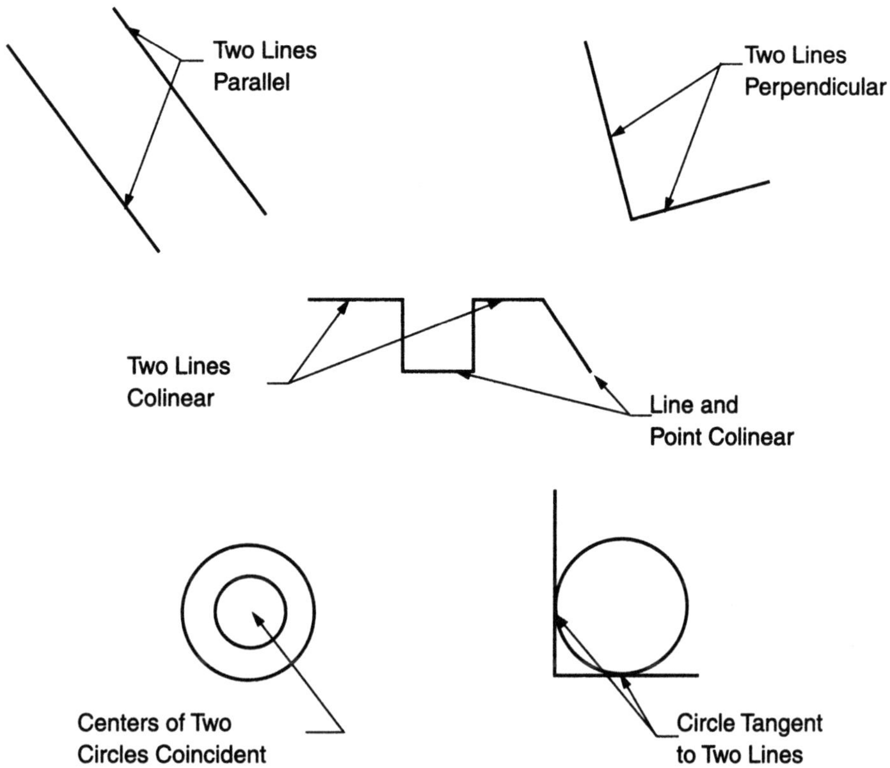

Figure 3–1 Types of Geometric Constraints

Tangent With this constraint, you establish a relationship between an arc (or a circle) and a line such that the two entities will always remain tangent to one another. If you create a pre-defined fillet on a sketch, chances are that this type of constraint was internally applied for you by the software.

Figure 3–2 shows a 2-D sketch that has geometric constraints applied to it as indicated next to the figure. Note that the shape of the geometry is now well defined. At this point, you can add ground and dimensional constraints (discussed subsequently) to the shape to define the exact sizes of the features that make up the geometry.

Guided Tour 3.1

- Start up Pro/ENGINEER and open a new file in *Sketch mode*. Turn on the Intent Manager by checking the open box near the top of the Menu Manager.

- In this Guided Tour, you are going to create a constrained 2-D geometry by using the Intent Manager. You will then apply additional constraints to achieve the desired results.

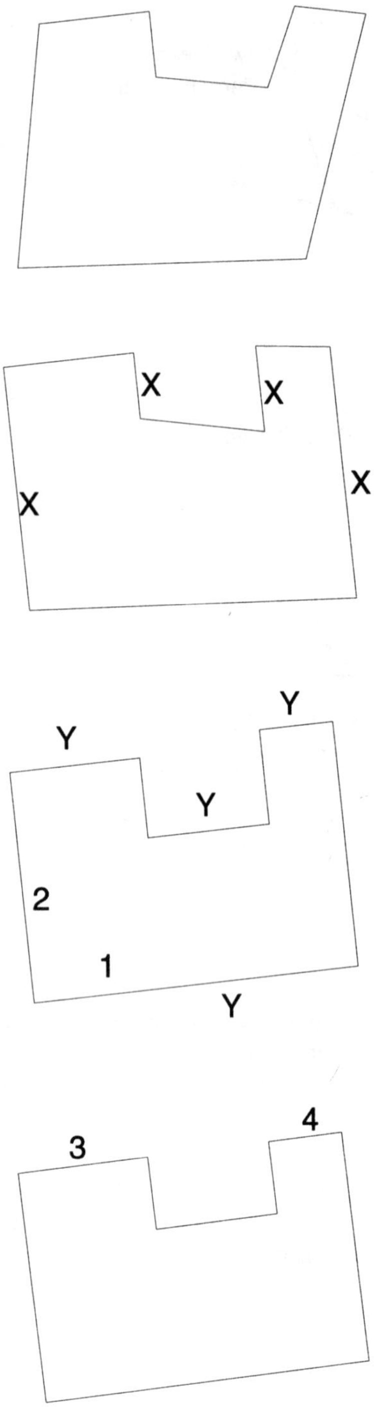

Original Sketch of 2-D Geometry

Parallel Constraints applied between lines labeled with an "X."

Parallel Constraints applied between lines labeled with a "Y." Perpendicular Constraint applied between lines 1 and 2.

Colinear Constraint applied between lines 3 and 4.

Figure 3–2 Application of Geometric Constraints to 2-D Shape

- Sketch the geometry shown below by first sketching the outline using the *Line* option. Note that since you have the Intent Manager on, your sketch will tend to snap lines to be vertical, parallel, perpendicular, etc. You may need to use the *Modify* and *Regenerate* commands to get the correct dimensions.

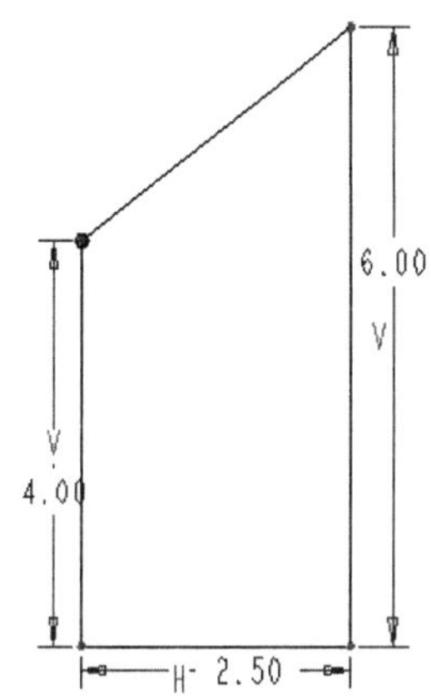

- Add a horizontal constraint to the top line by selecting *Constraints* from the Menu Manager. Note: your sketch may be overconstrained. If this is the case, you will be asked to delete one of your constraints. In the case, below the dimension with a value of 6 was deleted.

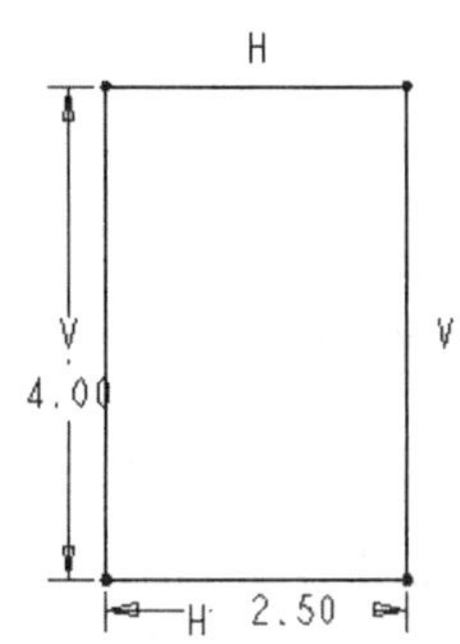

- Now sketch an arc across the bottom of the geometry as shown in the next figure.

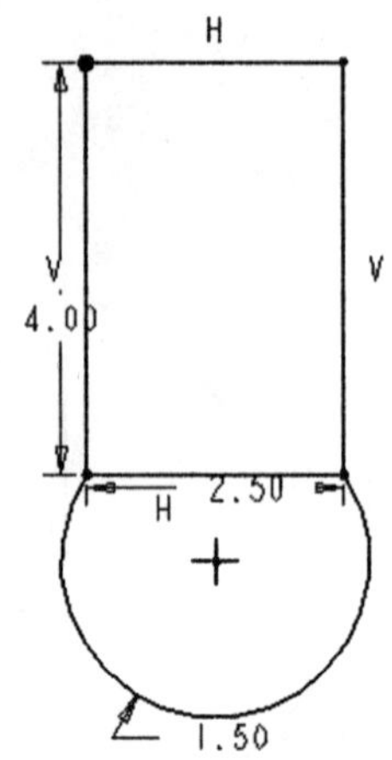

• Add tangential constraints between the arc and the two vertical sides.
• Your final sketch should be similar to the following figure:

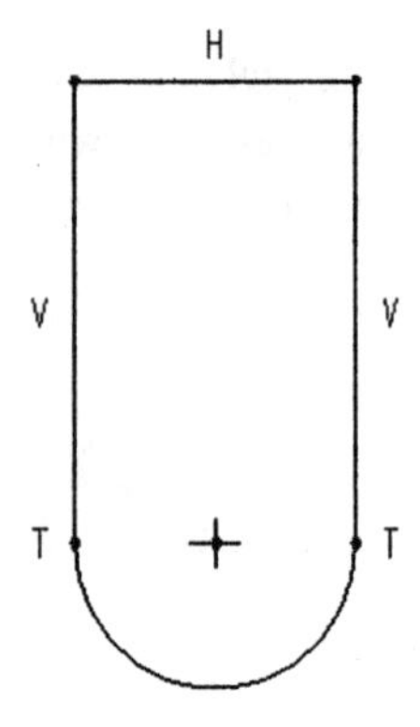

3.2.2 Dimensional Constraints

Dimensional constraints define the *size* of your 2-D shape. Sometimes when you apply this type of constraint, you will define a size that is inconsistent with other dimensions or inconsistent with the geometric constraints on your sketch. You may have run into this problem during Guided Tour 3.1.The type of dimensional constraints that are available to you include:

Linear This type of dimensional constraint is used to define the linear distance between two lines, between a line and a point or between two points. This type of constraint can be further subdivided into either vertical, horizontal or aligned dimensions. A vertical dimension is the vertical distance between two entities; a horizontal dimension is the horizontal distance between two entities; and an aligned dimension is the actual distance between two entities. Aligned dimensions can only be used to define the distance between two points.

Angular Angular dimensions are used to define the angle between two lines on your sketch. The lines between which the constraint is applied do not necessarily need to intersect on the sketch itself, however, you cannot define an angular dimension between two parallel lines.

Diametral/Radial Diametral constraints are used to control the diameter of circles on your sketch. Radial dimensions are also available to control the radius of any arcs on your sketch.

Figure 3–3 illustrates the type of dimensional constraints that you can use to define the size of your 2-D geometry.

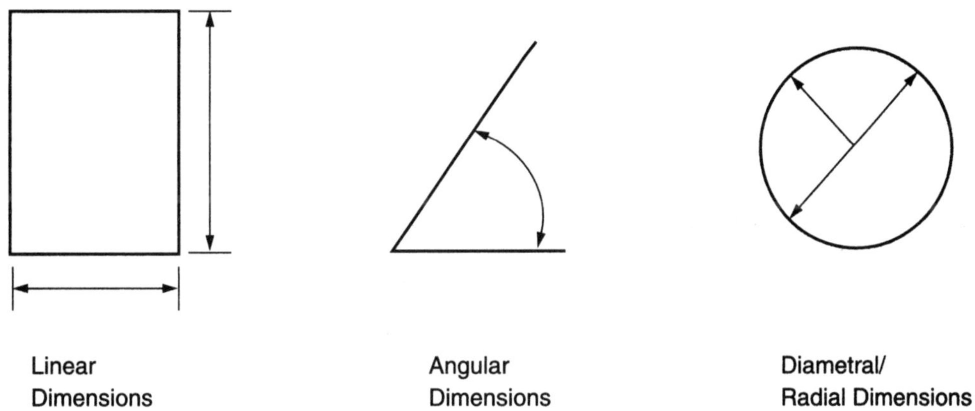

Figure 3–3 Types of Dimensional Constraints

Figure 3–4 shows a 2-D shape after the application of dimensional constraints. With the combination of geometric and dimensional constraints attached to the geometry, this shape is now fully constrained—there are zero degrees of freedom remaining. Note that if any of the dimensional constraints are modified, the geometric constraints will help maintain the *shape* of the geometry, but the *size* will change according to the newly input dimensional value.

3.2.3 Adding Dimensional Constraints in Pro/ENGINEER

Dimensional constraints in Pro/ENGINEER are applied during the Sketch stage of the geometry. This is done for you as you create your sketch with Intent Manager turned on, but it is also fairly easy to create your own dimensional constraints with Intent Manager turned off. There are three types of dimensional constraints we will discuss here: linear, angular, and radial. For linear dimensions, you can choose two lines, a line and a point, or two points to dimension between. After you have selected the two entities between which you would like to dimension, you will pick the location for the text. Place the text away from the drawing so that the geometry is visible. The geometric entities are selected using the left mouse button, and the text is placed using the middle mouse button. Angular dimensions are added by selecting the two nonparallel lines on the sketch you would like to constrain and then placing the text at an appropriate location. Radial dimensions are included by selecting the circle or arc for dimensioning and placing the text at the desired location. The *Autodimension* option in the Menu Manager is used to quickly add dimensions and constraints to a section.

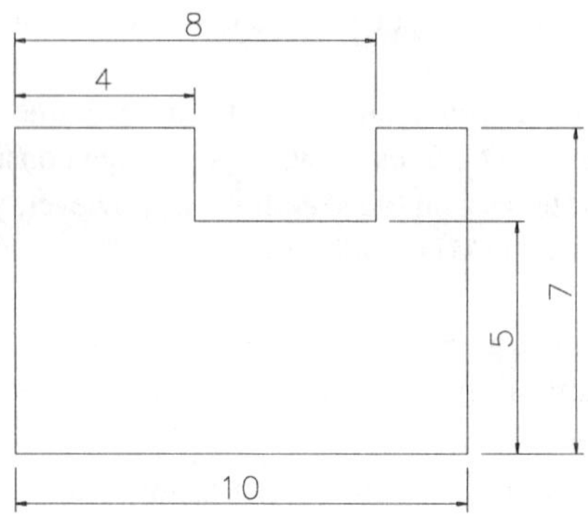

Figure 3–4 Two Dimensional Shape After Application of Dimensional Constraints

Guided Tour 3.2

- With Intent Manager turned off, carefully sketch the geometry in the following figure:

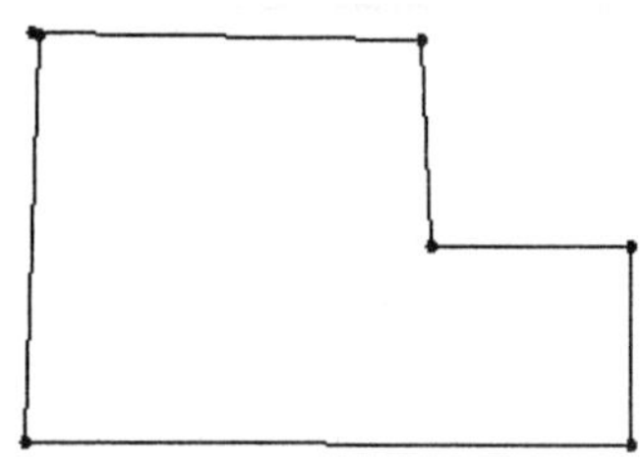

- Now add your dimensional constraints by selecting *Dimension* from the Menu Manager and clicking on the endpoint of the lines. Be careful not to overconstrain your sketch. Select *Regenerate* when you are finished. Your sketch should appear similar to the following figure (your values may be different):

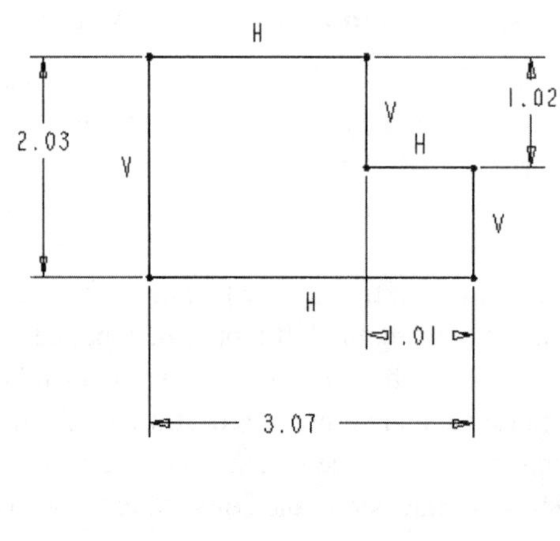

• Select *Modify* to change the values and then *Regenerate* until you obtain the final geometry shown below.

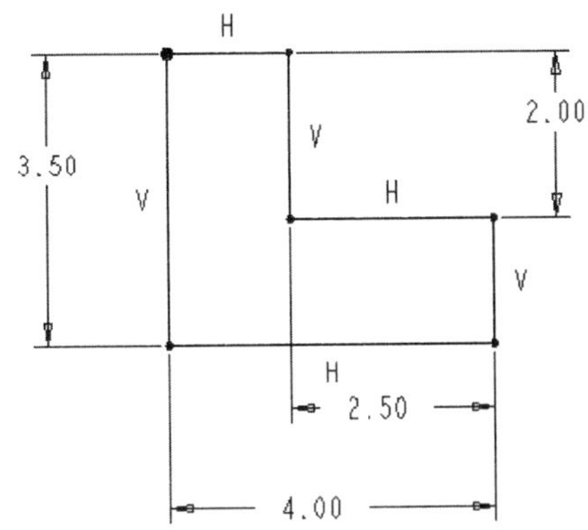

3.3 Modifying Dimensional Constraints to Size Your Sketch

When you are finished with your sketch, you will most likely want to change some of the dimensional constraints on your drawing so that your geometry is the correct size. You already have some practice at modifying the dimensional values from Guided Tour 3.2, but let's look at this a little more closely.

Typically, you can modify the dimensional values such that they are of a specific size or they are related to the size of another feature on the sketch. Dimensions that are a specific size are easily modified individually (i.e., you merely specify the numerical value for the dimension). With dimensions that are related to the size of other features on your sketch, you can establish equational relationships between them. For example, with the fully constrained shape shown in Figure 3–4, you could modify the dimensions such that the dimension with a value of "4" would always be half of the value of the dimension that is currently "8." If you change the value of the larger dimension, the smaller dimension will change accordingly. Figure 3–5a shows the original constrained geometry with the value of dimension A set to be half of the value of dimension B (the equation for this relationship is: A=0.5*B). Figure 3–5b shows the same geometry after changing the value of dimension B.

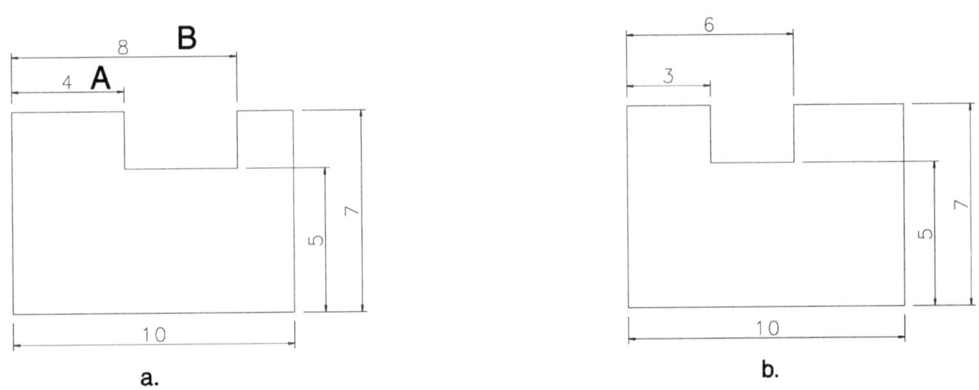

Figure 3–5 Modifying Dimensional Constraints Related by an Equation

When modifying dimensional constraints either individually or by equational relationships, you may find it easier to work from the "inside out" or from the "outside in," depending on your initial sketch. If you modify dimensions randomly, the screen may not always show the desired geometry, and the display can be confusing. For example, suppose you started with the geometry from Figure 3–6a (an equational relationship has *not* been added between A and B) and modified the dimensions so that C = 5 (shown in Figure 3–6b) and then B = 3 (shown in Figure 3–6c). If you then change dimension A so that it is half of dimension B, your original shape will be restored and maintained. However, in the intermediate steps, your sketch may be confusing to you. If, however, you had started with the smaller dimensions first and moved to the larger dimensions (from the *inside out*), you would have set A equal to one half of B first, changed B to 3, and then changed C to 5. The shape of your sketch would have been maintained throughout the entire modification procedure.

As with anything else in computer-aided design, it is important for you to think about what you are doing before you actually try to do it. A good rule of thumb to follow is: If you are modifying geometry by making it *bigger,* work from the outside in (overall dimensions first), and if you are modifying geometry to make it *smaller,* work from the inside out (overall dimensions last). It is also generally best to start by applying equational relationships to your sketch as appropriate.

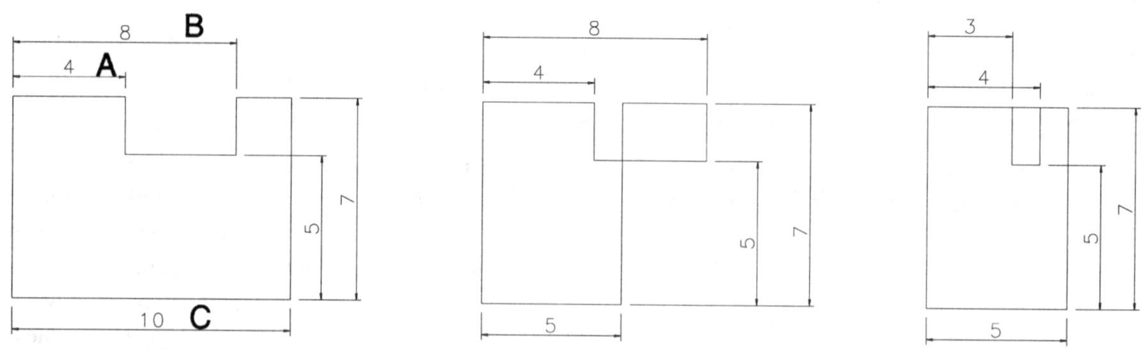

a. b. c.

Figure 3–6 Modifying Dimensional Constraints in Random Fashion

3.3.1 Modifying Dimensional Constraints in Pro/ENGINEER

You will select *Modify* from the Menu Manager to change the value of any of the dimensional constraints on your sketch. When you select the *Modify* option, you must select the entity you wish to modify. A dialog box will appear at the command line along the bottom of your screen, asking you to enter the new value. Once entered, the value will not change until you select *Regenerate* from the Menu Manager. Because the new values will not automatically update, you can change several at a time. For large and complex geometries, it is a good idea to make several small changes, slowly working your way to the desired size and shape (several small changes help avoid the problems found in Figure 3–6). We can use this method to select and change each individual dimension, however, it may be beneficial to develop relationships between dimensions such that when we change one value, several others will change proportionally. This is called adding dimensional relationships.

3.3.2 Adding Dimensional Relationships in Pro/ENGINEER

Many software packages have a feature to program dimensional relationships. In Pro/ENGINEER, this is easily done using the *Relation* command found in the Menu Manager. Once the *Relation* option is selected, all the dimensional values change from numbers to variables, such as *sd1, sd2, sd3,* etc. Equations can then be entered in the dialog box found on the command line at the bottom of the screen. To create a new relationship between variables simply select *Add* from the Relation Menu Manager. Try Guided Tour 3.3 to see how this works.

Guided Tour 3.3

- Regenerate your part from Guided Tour 3.2 and select *Relation* from the Menu Manager. (Your variables should have different names. Use the variable names on *your* sketch to create the relationship.)

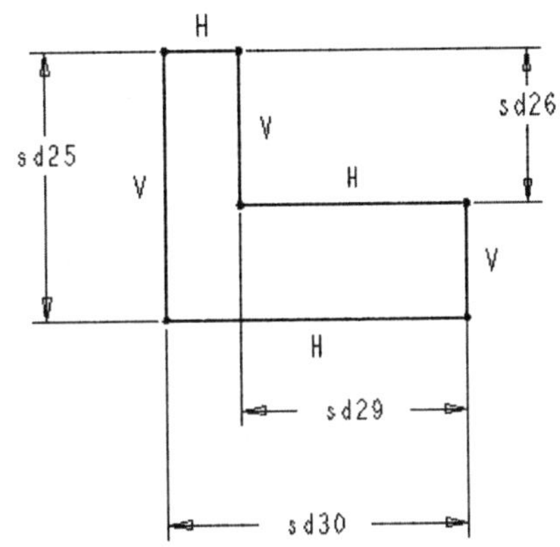

- Select *Add* from the Relation Menu Manager, and type the following relationships (using the correct corresponding variables):

sd29=2/3*sd30

sd25=3/4*sd30

sd26=1/2*sd30

- Hit *Return* twice and select *Regenerate*. Your sketch should look like the following:

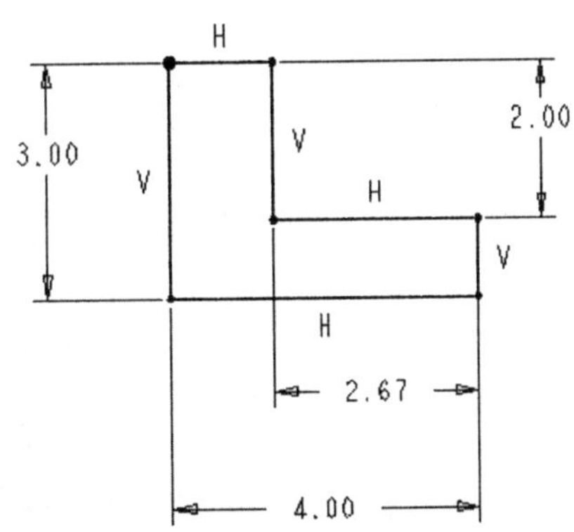

- Change the value *4* to *8* (use *Modify* and *Regenerate*). All the dimensions should change proportionally.

3.4 Constraining 2-D Geometry Automatically as You Sketch

Most 3-D solid modeling packages available today have a sketching aid built in to the software. If you start to draw a line that is nearly vertical, the software will "sense" that you want to draw a vertical line, and the cursor will snap to a location on the screen to make the line vertical. Similarly, if you begin to draw a line that is nearly parallel to (or perpendicular to) another line on the screen, the cursor will snap to a location so that the line is drawn parallel (or perpendicular). Some software packages may also automatically dimension your sketch for you as you are creating it. If this is the case with the software you are using, many of the constraints that you will need to apply to your sketch may already be in place when you finish.

Some of the constraints that the computer chooses to apply may not be appropriate for your design. If this is the case, you will need to delete the constraints that are automatically created by the computer and supply your own constraints in order to preserve your design intent.

3.4.1 Automatically Constraining as You Sketch in Pro/ENGINEER

In many cases, geometric constraints will be automatically applied for you by the Intent Manager as you sketch in Pro/ENGINEER. (Refer back to Chapter 2 for a discussion of the Intent Manager.) Sometimes you will need to delete constraints or add other constraints to your sketches because the Intent Manager is only making "best guesses" about your design intent. As you work, pictorial representations of the applied geometric constraints will be shown on your sketch so that you can tell what type of constraint the Intent Manager is applying. Figure 3–7 shows the graphic symbols associated with each type of geometric constraint in Pro/ENGINEER.

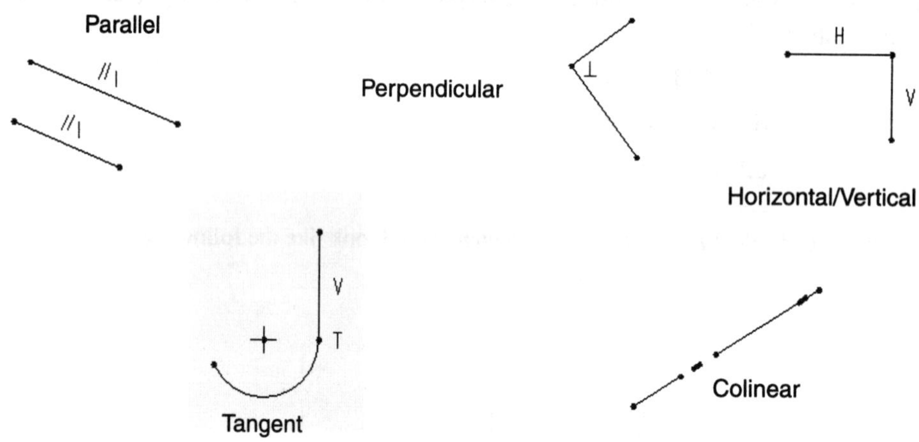

Figure 3–7 Geometric Constraint Symbols

Dimensional constraints will also be applied by the Intent Manager as you sketch in Pro/ENGINEER. The Intent Manager will identify points such as midpoints and center points of existing geometry. These points will assist you in creating your sketch so that it looks proportionally correct, however, the Intent Manager may *not* apply constraints for these relationships. If you want the relations maintained, you must identify them as equational relationships by using the *Modify* icon.

If the Intent Manager adds a constraint (either dimensional or geometric) that you do not wish to maintain on your final sketch, you must first delete the existing constraint before you can apply the one you want. You can delete dimensional constraints by picking the text of the constraint on your sketch.

Guided Tour 3.4

- Start a new geometry in *Sketch* mode.

- With the aid of the Intent Manager, sketch the following shape, making sure that the indicated constraints are part of the definition of the sketch. Start by sketching the vertical line. Finish by sketching the tangent line from the end of the arc to the tangent point on the circle. Note the parallel and perpendicular constraints.

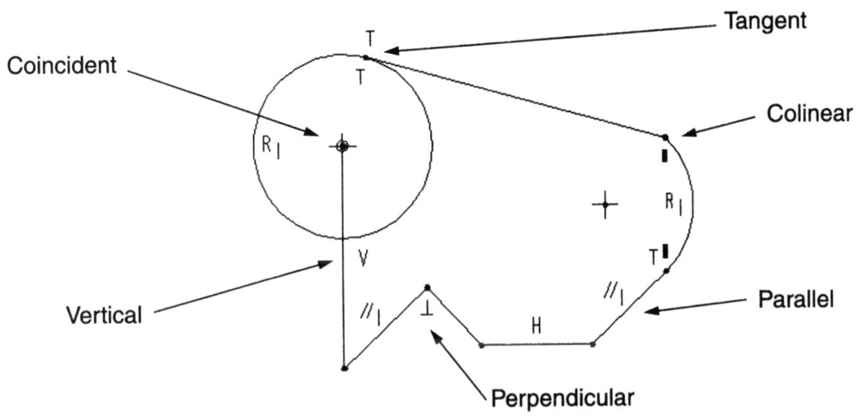

- Add a tangent constraint to the top of the arc (You may need to delete other constraints in order to do this.)

3.5 Strategy for Constraining Sketches

When applying constraints, it is important for you to *think* about what you are trying to accomplish. You must also think about the features of the part you may wish to change at a later date. If you think that you may want to change something later on, it is important that you apply a constraint to your sketch in such a way that you are able to easily make that change later on. For example, if you *know* that two lines on your sketch should always remain perpendicular to one another, you should apply a perpendicular constraint between the two lines to preserve this relationship. If, however, you *think* that the two lines should be perpendicular to one another, and you initially sketch them in that configuration, it might be better to apply an angular dimension as the constraint between the two lines. The initial value of the angular dimension could be 90 degrees (perpendicular), and if you determine later on that the two lines should meet at 85 degrees, you can easily make this change. If you do not apply an angular dimension initially but rather as a perpendicular constraint, you will need to delete the perpendicular constraint, apply an angular dimension, and then modify the angular dimension to be 85 degrees. You can save yourself time "down the road" if you *think* about how to apply constraints from the outset.

When sketching and constraining, you should first plan how you are going to sketch the shape and which constraints you would like to apply to the shape. In Pro/ENGINEER, the Intent Manager will assist you by adding dimensional and geometric constraints as you sketch, however, the dimensional constraints added by the Intent Manager will often need to be modified or deleted and new ones added in their place to preserve your design intent.

A common shape that occurs in engineered parts is the classic I-shape. Design intent for this shape typically maintains that the two flanges be of equal thickness and that the web be centered within the flanges. If this were sketched in Pro/ENGINEER, the constrained shape shown in Figure 3–8a might result (the results could vary depending on how it was sketched). It should be noted that the shape in 3–8a is fully constrained, however, with the dimensions that were automatically applied, it will probably be difficult to maintain the design intent if modifications are made to the dimensional values. If you wanted to change the thickness of one flange, you would need to change all three vertical dimensions to preserve the design intent. To constrain this shape so that design intent is easily preserved, it would be best to delete the vertical dimension of 70 and add a dimension controlling the height of the lower flange, setting it equal to the height of the upper flange. The design intent for centering the web could also be maintained by deleting the dimension of 52 and adding a dimension for the left side equal to the dimension of 30 on the right. The constrained shape that maintains design intent is shown in Figure 3–8b.

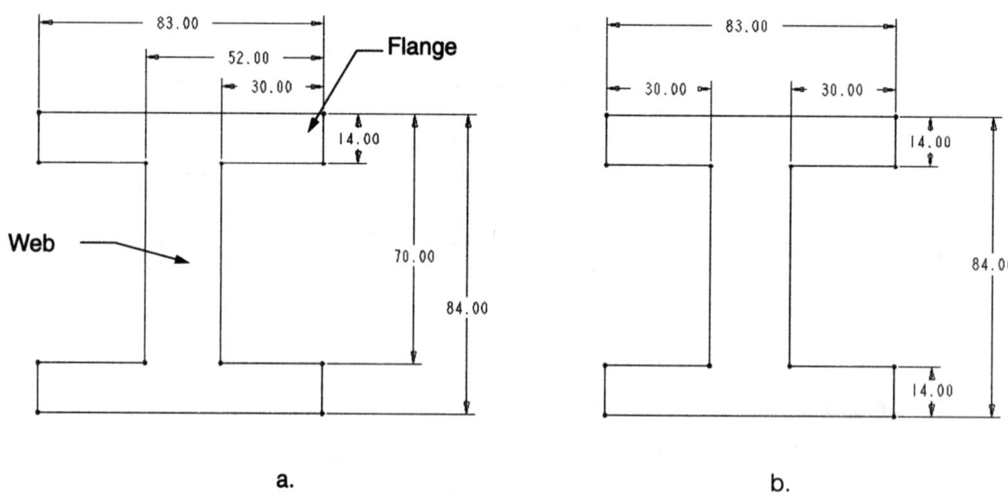

Figure 3–8 Constrained Sketch of I-Shape

When creating fillets around the end of a rectangle (seen in the following sketch) make sure that the fillet radius is not equal to half the width of the rectangle or you will end up with a self-intersecting profile that cannot be extruded or revolved. You can either make the fillet radius slightly smaller than half the width of the rectangle, or you can create a circle and trim it and the lines in that region. This is illustrated in Figure 3–9.

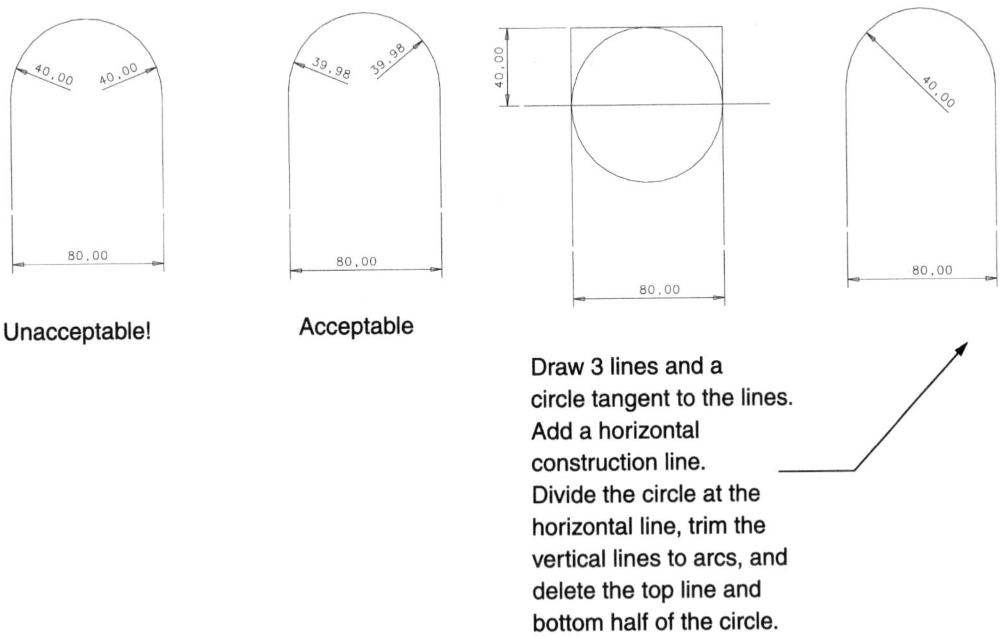

Figure 3–9 Creating a Round End Between Parallel Lines

An alternative way to construct this geometry is to draw two vertical lines and connect them with an arc as illustrated in Figure 3–10. Note that if you utilize this construction procedure, the end result may be "higher" than the original lines, and the length of the lines may need to be adjusted accordingly as you sketch it.

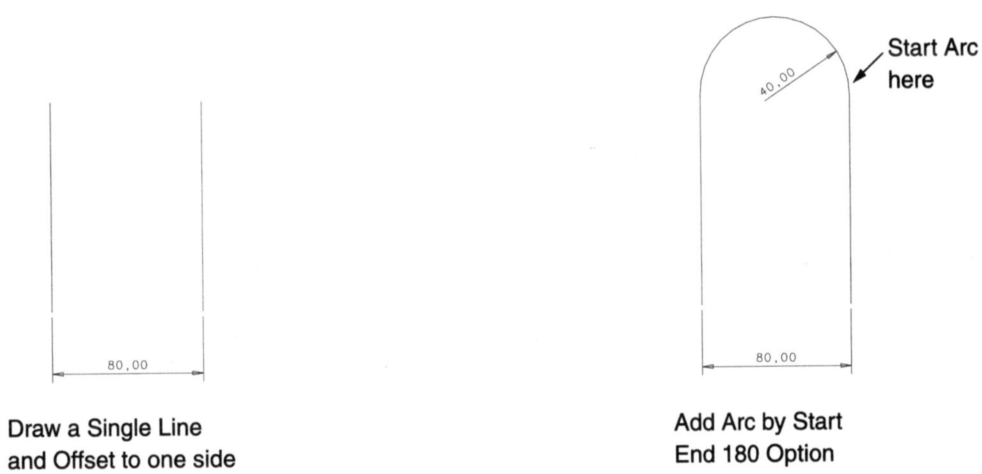

Figure 3–10 Alternate Method for Creating a Round End Between Parallel Lines

3.6 Troubleshooting

Problem	Potential Solution
You have modified a dimension and your sketch no longer looks the way you want it to.	In order to preserve your design intent, you may need to delete some of the dimensions applied by Pro/ENGINEER and add new ones in their place.
You have modified a dimension and your sketch became "goofed up" following a regeneration.	Select *Undo* or *Unregenerate* to restore your original sketch. Check the way that dimensions are applied on your sketch and delete or add, as needed, before attempting to change the dimension again.
Your 2-D geometry does not start and stop at the same point.	Apply a *Coincident* constraint at the corner to force two lines to share a common endpoint.
You are having trouble selecting a geometric constraint to delete it.	After selecting *Delete*, choose *Query Select*. This will give you several items to choose from.
You are trying to dimension between two "short" lines and Pro/ENGINEER keeps picking the endpoints of the lines instead of the lines themselves.	Point to a location slightly off the line to highlight the line and not its endpoints for selection.

3.7 Chapter Summary

In this chapter, you first learned about constraints and how they are useful in designing a flexible model. The types of constraints considered were geometric and dimensional. You then learned about modifying constraints to size your sketch and establishing equational relationships between dimensions on your sketch. You learned that many constraints are automatically added to your geometry as you sketch. You also learned the necessity of sometimes deleting the automatically applied constraints in order to preserve design intent. Key Words for the chapter are tabulated in the following:

Angular Dimensions	Geometric Constraints
Coincident Constraint	Linear Dimensions
Colinear Constraint	Match Dimensions
Constrain & Dimension	Modify
Diametral/Radial Dimensions	Parallel Constraint
Dimension	Parametric Modeling
Dimensional Constraints	Perpendicular Constraint
Intent Manager	Tangent Constraint
Equational Relationship	Variational Geometry

Exercises

On a clean sheet of paper, develop a **Plan of Action** for the object assigned by your instructor. Describe the steps you would follow to create the object in Pro/ENGINEER. Use sketches as appropriate in your Plan of Action. Include the constraints you would apply in creating the object. Then, create the object in Pro/ENGINEER, making sure it is created to the correct *size* as well as the correct *shape*. Object features in the following exercises are symmetrical unless otherwise noted. Obtain a view of the object specified by your instructor, and turn in a hard copy as appropriate.

1. Sketch the following figure. Apply a perpendicular constraint between line A and B. Apply the same constraint between A and C. What happens when you apply a vertical constraint to Line B? Label each line and describe all the constraints need to transform this object into the shape of an "I".

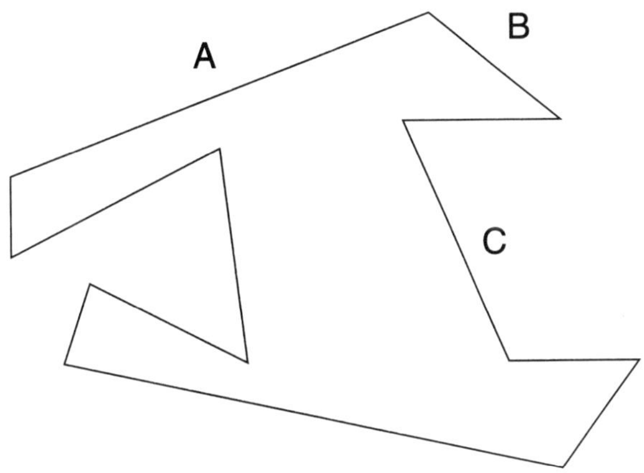

2. Sketch the following figure. Apply the minimum number of constraints to form a normal slot. List and label all constraint locations.

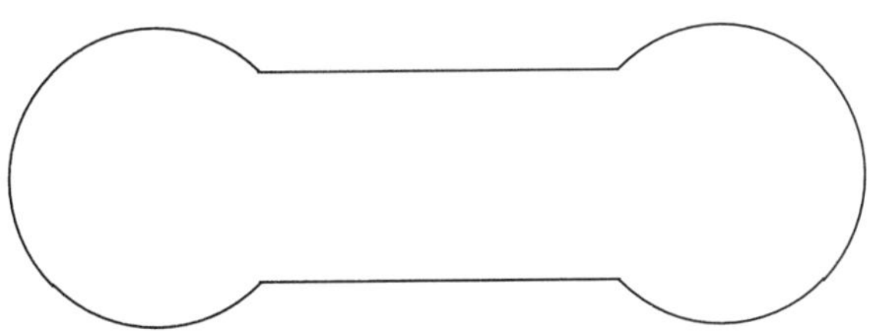

3. Sketch the following figure. Add the dimensional relationships setting A = 1 and B = 1. Regenerate and set A = 2 and B = 3. What happens?

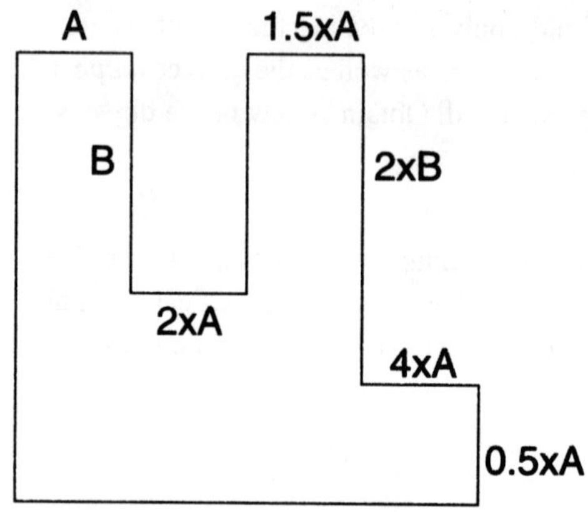

4. Sketch the following figure Add the dimensional relationships setting A = 2 and B = 1. Regenerate and set A = 1 and B = 3. What happens?

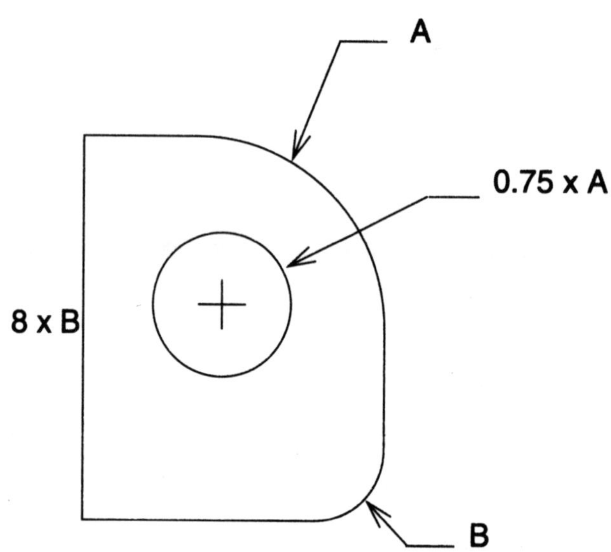

CHAPTER 4

Creating 3-D Parts from 2-D Geometry

Some of the most common modeling techniques used in creating 3-D parts involve first sketching or drawing 2-D geometry and then extruding or revolving the geometry to form a solid. If you are familiar with the Play-Doh FunFactor™, you can understand the basic concept of extrusion. Extrusion is a material-forming process which involves forcing a material through a well-defined opening to create a final product that is in the shape of the opening. When you play with the FunFactory™, you form the Play-Doh in the shape of stars, tubes, spaghetti, and so on by extrusion. In computer modeling, you define the shape for the extrusion by sketching the two-dimensional geometry, and then you "push" that shape back a given distance to form a solid. Figure 4–1 shows a two-dimensional sketch and the resulting 3-D part that was created by extruding the shape.

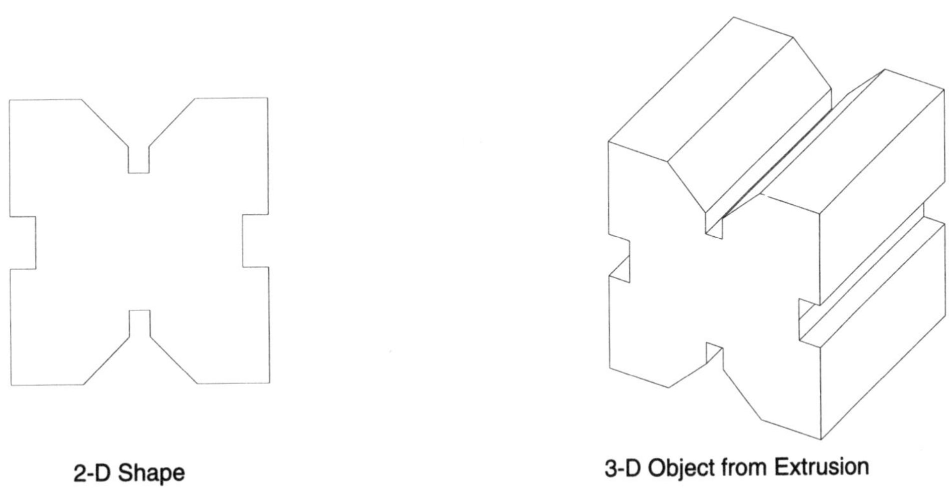

Figure 4–1 Object Created by Extrusion

Revolution is another technique used to create a 3-D part from two-dimensional geometry. With this technique, a given 2-D shape is revolved about an axis. As the shape is revolved, a solid is swept. Figure 4–2 includes a 2-D shape that was revolved about the indicated axis to create the 3-D part shown on the right.

Extrusion and revolution are powerful methods for creating parts by Computer Aided Design techniques. Each method will be discussed in more detail in subsequent sections of this chapter. However, in order for you to effectively use these techniques, it is important that you understand how the 2-D geometry is sketched or drawn. Sections 4.1–4.3 discuss the creation of 2-D sketches, and the remaining sections are devoted to extrusion and revolution as well as other topics related to 3-D sketching.

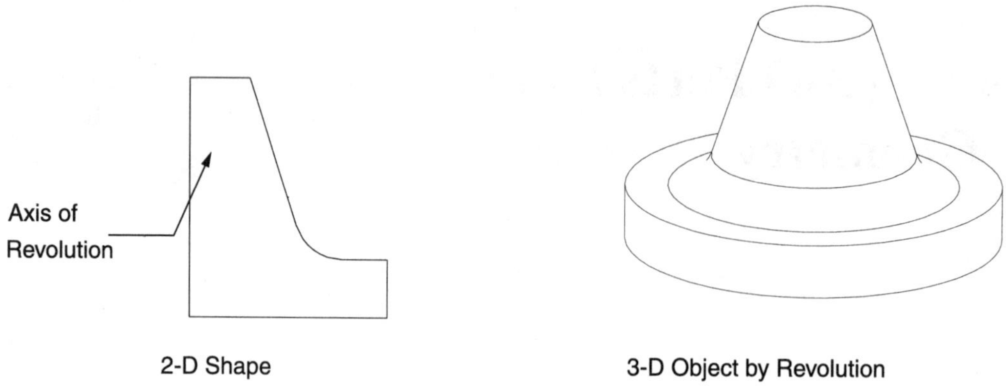

Figure 4–2 Object Created by Revolution

4.1 Sketching or Drawing 2-D Geometry

Two-dimensional sketching is accomplished on a *plane* in three-dimensional space. Most modern software has a means of helping you align things properly as you sketch so that your final shape looks "close" to what you desire. This means that if you sketch a line close to vertical, then the software will automatically "force" it to the vertical position. Other times, this means that a grid is included in the sketching area to help you align things as you sketch. After you have finished your sketch, you can apply constraints (described in Chapter 3) to achieve the desired final 2-D shape. The basic two-dimensional geometric entities that can be combined to develop your sketch include the following:

Points A point is a zero-dimensional geometric entity. It has no length or width, but it defines a location on the sketching plane exactly. Points are typically added to sketches for reference purposes—the point can be used to locate other geometric entities more accurately.

Lines A line is a one-dimensional geometric entity defined by two points. It has a length but no width or thickness. In your math courses, you probably learned that lines are infinite in length, however, when you are creating lines in solid modeling software, they are assumed to be of fixed length. To create a line, you specify its two endpoints. With computer software, lines can also typically be drawn that are parallel to or perpendicular to any other line that already exists on the drawing.

Polylines Polylines are a convenient way to construct several connected lines on your sketch. With polylines, the endpoint of the first line drawn becomes the starting point for the second line, the endpoint for the second line becomes the starting point for the third line, and so on. In this way, you define two points for the first line but only one point for each subsequent line in the sketch.

Circles A circle is defined by any three non-colinear points on the sketching plane. When creating circles using computer software, there are typically several methods available depending on how you want to create them. You can choose to define any three points on the circle, or you can define two points on the diameter of the circle. Other methods include specifying the location of the circle center and supplying either the radius or the diameter of the circle or creating a circle tangent to two or three different lines on your sketch.

Arcs Arcs are portions of circles that sweep angles less than 360 degrees. To create an arc by sketching, you typically will employ methods similar to those used in creating circles. You can create and arc by specifying three points on the arc, the start, end, and the radius, etc.

Splines A spline is a general curve. It takes a minimum of four points to define a spline. In solid modeling software splines are specified by selecting the points that make up the spline. The computer will then fit the spline through those points.

Polygons With most software packages you have the ability to create basic polygons. Typically, you can create polygons that have many equal sides. Rectangles are among the those polygons that are commonly created. To create rectangles, you will usually specify two diagonal corners or the location of the center of the rectangle and one corner.

Figure 4–3 shows the various types of two-dimensional geometric entities that are typically created in making sketches with solid modeling software.

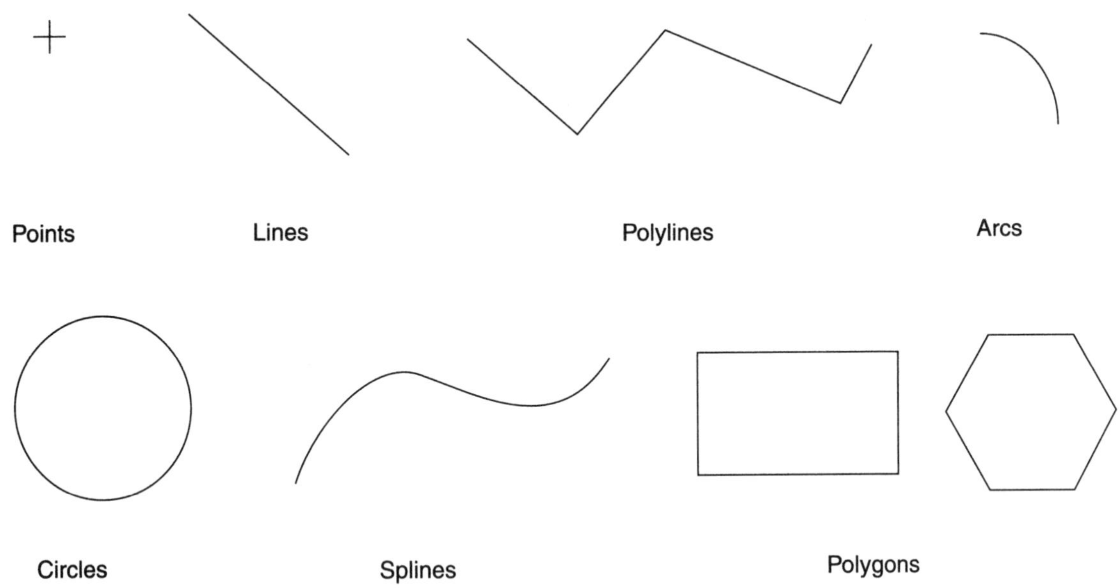

Figure 4–3 Two-Dimensional Geometric Entities

When creating 2-D shapes, you will need to specify the points that define the geometric entities. For example, a line needs two points on the working plane for its definition. Points are often specified by merely clicking the mouse button at the desired locations on screen. Alternatively, there may be **implicit** or **explicit** point specification methods available with a software package. Explicit point specification methods include providing the x- and y-coordinates of the point relative to the origin or to some other reference point. Implicit point specification methods are used to designate a point location relative to geometric entities that already exist on the sketch. With implicit point specification methods, you select the *end* of a line, the *midpoint* of a line, the *center* of a circle, the *intersection* of two lines, and so on.

4.1.1 Sketching 2-D Geometry in Pro/ENGINEER

Pro/ENGINEER software is based on the premise of "shape then size." What this means is that you should sketch the *shape* of your geometry so that your design intent is defined. After you have correctly sketched the shape of your geometry, you then apply exact dimensional values for your sketch so that the *size* is appropriate for the functionality of the design. In this section, you will learn the basics of sketching in Pro/ENGINEER as well as how to use your sketches to create 3-D parts by extrusion or revolution. Later in this chapter, you will learn how to size your geometry appropriately.

4.1.2 Automatic Constraints

As you are making your 2-D profiles, Pro/ENGINEER will assist you in creating your sketches. Pro/ENGINEER tries to "guess" your design intent as you sketch. For example, if you draw a line that is close to vertical, the software guesses that you really want to draw a vertical line, and it will add a vertical constraint prior to the regeneration of your sketch. Similarly, if the software thinks that you want to draw a line parallel to another line that is already on your drawing, it will add a parallel constraint to that line, and after regeneration, it will be drawn parallel.

4.1.3 Sketching 2-D Geometry

There are several options available to you as you create your 2-D sketches in Pro/ENGINEER. The commands that you will use to make your sketches are described in the following list of (All of these commands are found under the Sketch option in the Menu Manager):

Lines Use this option to draw single lines or polylines on the screen. You will be prompted to locate the start for the first line and to locate the ends for subsequent lines. To start or end a line, click with the left mouse button. To stop drawing lines, use the middle mouse button.

Rectangle This command is use to generate simple rectangles. You click with the left mouse button to define one corner and click again to define the other.

Arc At this selection, there are three different options you can use to sketch circular arcs: 1) Tangent End—click near the end of a line and select an endpoint to sketch the arc, 2) Concentric—click the center point of an existing arc or circle, 3) 3 Tangent—draw a circle tangent to three lines on your sketch, 4) Fillet—click two lines to create a circular fillet, and 5) Center/Ends—click the circular center of the arc, then click the start and end points defining the arc.

Circles At this selection, there are five different options you can use to sketch circles: 1) Center/Point—click the center point and then a radial point on the circle, 2) Concentric—click the center point of an existing arc or circle, 3) 3 Tangent—draw a circle tangent to three lines on your sketch, 4) Fillet—click two lines to create a circular fillet, and 5) Three Points—click any three non-colinear points to define the circle.

Splines The command to create a spline is found under *Advance Geometry.* Use this option to create a spline on your sketch by clicking the points that define it. Under *Advance Geometry* you can also select conic sections, elliptic fillets, and text—all of which can become geometric features of your model.

4.2 Fillets and Rounds

Fillets (the "t" is pronounced) are geometric entities that are commonly found on engineered parts. Therefore you will often draw fillets on your sketches as you create your parts. A fillet is a special kind of geometric entity that consists of an arc with a specified radius that is tangent to two lines on your sketch. Fillets are typically used to "round off" a square corner. Rounds are similar to fillets except that they occur on an "outside" corner whereas fillets occur at "inside" corners. Figure 4–4a shows a 2-D sketch, and Figure 4–4b shows the resulting geometry after a fillet and a round have replaced two of the sharp corners on the sketch. Note that you could create a fillet or a round by first creating a circle tangent to the two lines in question and then trimming both the circle and the lines to their points of tangency. Fortunately, most modern software will automatically create the arc and perform all the trimming operations for you when you specify that you want to create a fillet on your sketch.

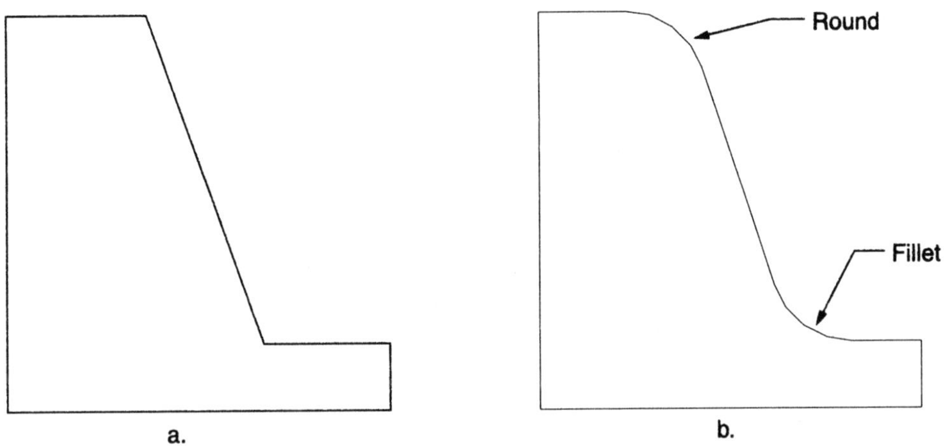

Figure 4–4 Two-Dimensional Fillets and Rounds

4.2.1 Fillets and Rounds in Pro/ENGINEER

You create fillets and rounds at corners or between two curves on your sketch by choosing *Arc* then *Fillet* under the Sketch Menu Manager. It is important to note that Pro/ENGINEER does not make a distinction between a fillet and a round but uses the term "fillet" to mean *both* fillets and rounds. When this option is selected, you will be prompted to select the two lines that will intersect to make the resulting round. The radius of the round is based on one of the points you selected on the lines. The closer of the two points from the intersection will be selected as the radius of the fillet.

Guided Tour 4.1

- Start up Pro/ENGINEER, and from the Menu Manager select: *Create—Solid—Protrusion—Extrude—Solid—Done*. This sets you up to create another extruded part. Select the *Line* command and create a sketch that nearly fills the main graphic area similar to the following one (Note: to show the constraints, click *Constraints* from the Menu Manager):

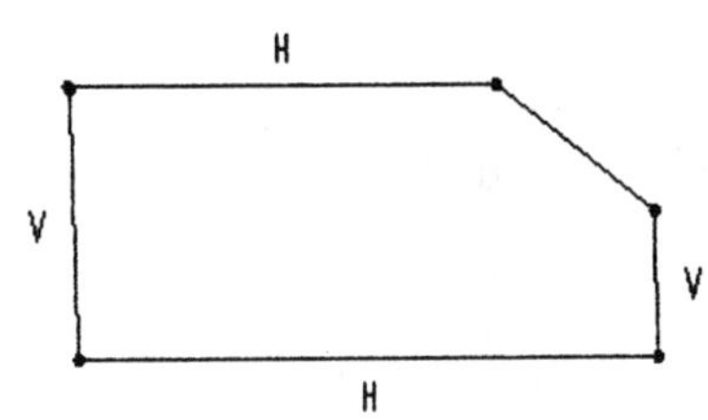

- Select *Rectangle* from the Sketch Menu Manager and create a rectangle in the middle of the figure just drawn. At the right end of this rectangle, draw a circle by the *Center/Point* option that goes through the corners of the rectangle (select the midpoint of the line as closely as possible). At the left end of the rectangle, draw an arc using the *Center/Ends* option. When you are finished, your sketch should be similar to the following (again, select *Constraints* to view the automatically placed constraints):

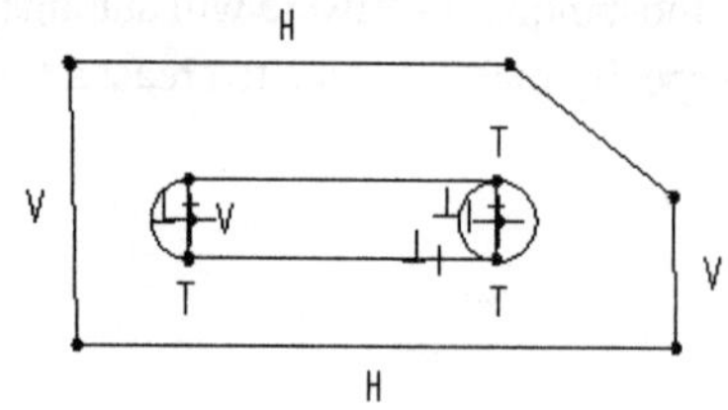

- Add fillets to the three right angles of your sketch, letting Pro/ENGINEER choose the size of the fillet for you. When you are finished, your sketch should look similar to the one shown here:

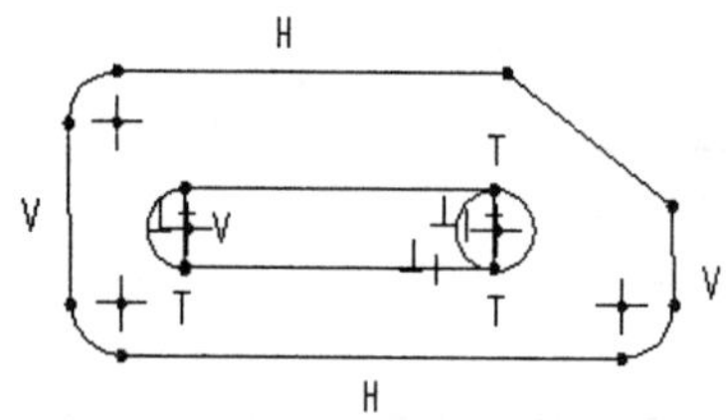

4.2.2 Using the Pro/ENGINEER Mouse Sketch

One option that you may find useful in creating your sketches is the **Mouse Sketch** command. This is a time saving feature which allows you to toggle between sketching lines, arcs, and circles without the use of the Menu Manager. Table 4-1 shows the "click strokes" required to use Mouse Sketch.

Table 4-1 Pro/ENGINEER Mouse Sketch Command Modes

Feature	Left Mouse Click	Middle Mouse Click	Right Mouse Click
Line	Start–Finish	Abort	
Circle	Abort	Start–Finish	
Tangent Arc		Abort	Start–Finish

You to build your sketch more efficiently by using Mouse Sketch than by selecting commands from the Menu Manager. It is strongly suggested that you practice using this feature of Pro/ENGINEER.

4.2.3 Deleting Geometry from your Sketch

When in the process of creating 2-D geometry in Pro/ENGINEER you will often need to remove items from the screen. *Delete* is the command used to delete entities in Pro/ENGINEER. Note that when you delete entities in Pro/ENGINEER, you will NOT be prompted to indicate whether or not you really wish to delete the selected entity before it is removed from the screen. If you selected the wrong item to delete, use the *Undelete Last* command to restore it. We will revisit the delete function again once we have created our three-dimensional object.

4.3 Trimming and Extending Entities

As you are creating 2-D entities on the screen you may need to modify them in order to end up with the desired final shape. Two of the more useful operations include **Trim** and **Extend**. With *Trim*, you are removing part of a given entity to its *point of intersection* with another entity on the screen. For example, Figure 4–5a shows a line and a circle. If the right end of the line is trimmed to the circle, the shape shown in Figure 4–5b results. If the left end of the line is also trimmed to the circle, then the shape shown in Figure 4–5c results. Similarly, if the upper portion of the circle is now trimmed to the line, the shape shown in Figure 4–5d results. Typically there are software-specific peculiarities involved in trimming circles. Some software packages trim circles clockwise, some counterclockwise. Sometimes packages trim circles from the three-o'clock position, but others do not. It is best that you experiment with the particular software package you are using to learn exactly how it trims circles.

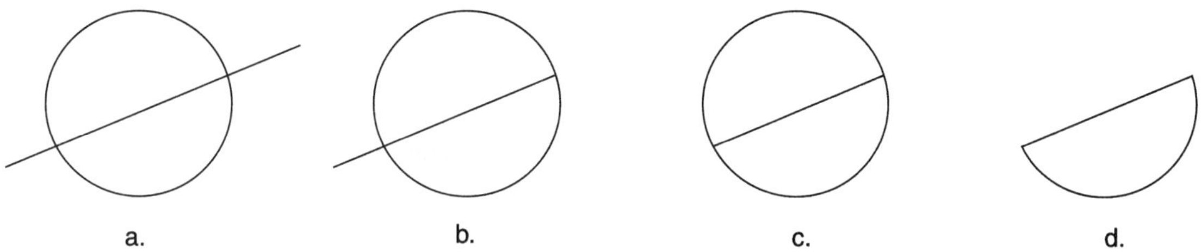

Figure 4–5 Trimming Operations

Extend is used to lengthen geometric entities to their point of intersection with another entity on your sketch. Sometimes the point to which a line will be extended is an "implied" point of intersection. Figure 4–6a shows two lines. Note that these two lines do not currently intersect. Figure 4–6b shows the same two lines after the line labeled "A" has been extended to its implied point of intersection with line B. Figure 4–6c shows these lines after line B has been extended to its point of intersection with line A.

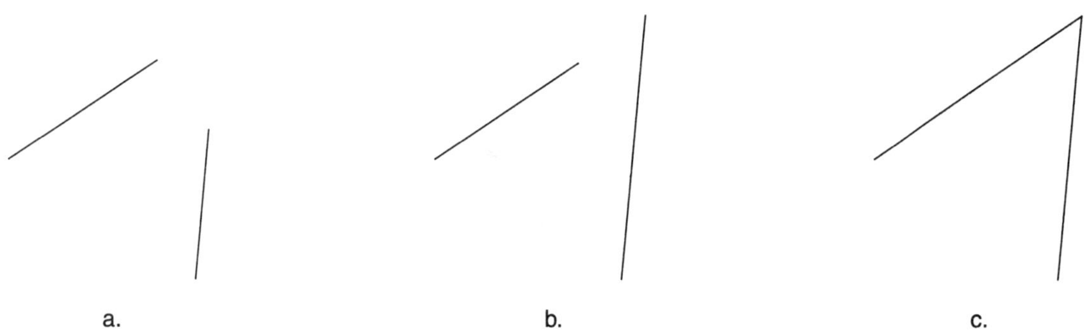

Figure 4–6 Extend Operation

4.3.1 Trimming and Extending in Pro/ENGINEER

Trimming and extending geometry in Pro/ENGINEER is accomplished by first selecting the Geometry Tools (*Geom Tools*) command in the Menu Manager and then selecting *Trim*. Both trimming and extending are accomplished with the use of this command. For most trimming and extending operation the *Bound* and *Corner* commands will be used.

When trimming or extending lines or curves in Pro/ENGINEER, you should select the line or curve near the end that you wish to affect. For the lines shown in Figure 4–7a, you would use the *Trim—Bound* mode and select the boundary as line B, then the line to trim. If you select line A near the left end, the geometry shown in Figure 4–7b will result. If when prompted for the line to trim, you cursor-select line A near the right end, the geometry shown in Figure 4–7c would result.

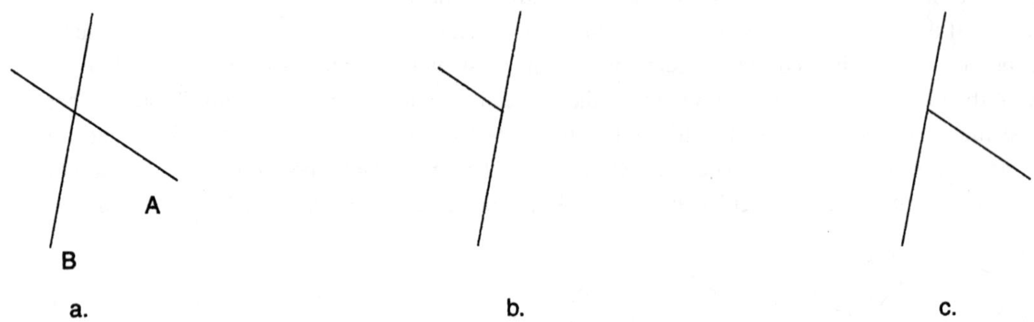

Figure 4–7 Trimming Lines in Pro/ENGINEER (Bound command)

Trimming circles in Pro/ENGINEER requires a bit more practice than trimming lines or arcs. A circle must first be divided into several arcs before the arcs can be trimmed or deleted. To trim a circle bounded by a line (Figure 4–8a), we first select the *Divide* command from Geometry Tools then click on the part of the circle you wish to remove. You should place at least two divides on the circle, similar to Figure 4–8b. After the *Trim* command, select *Bound* and click the line and the part of the arc you wish to keep. Repeat this, and you will end up with a sketch similar to Figure 4–8c. The final step is to delete the unwanted arc. Alternatively, you may find it easier to first intersect a circle into arcs. To intersect the circle, choose the *Intersect* command from the Geometry Tool Menu Manager and select the line and circle. Then simply delete the unwanted arc.

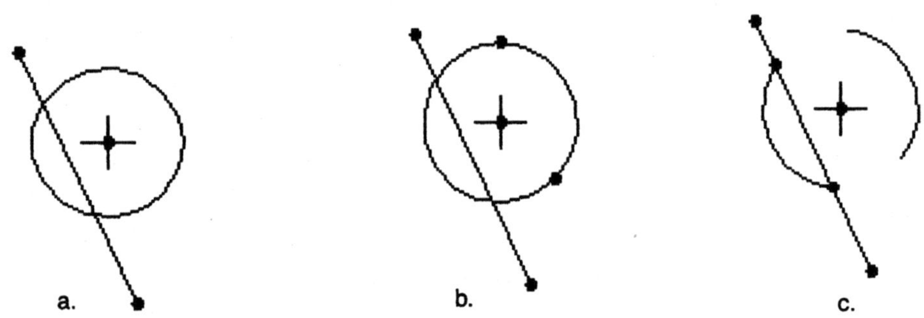

Figure 4–8 Trimming Circles in Pro/ENGINEER

Guided Tour 4.2

- Starting with the sketch you developed in Guided Tour 4.1, offset the angle line in the upper right corner in a direction away from the figure. Again, let Pro/ENGINEER choose the offset distance. You may need to flip the side for the offset in order to achieve this. When finished, your sketch should look similar to the following:

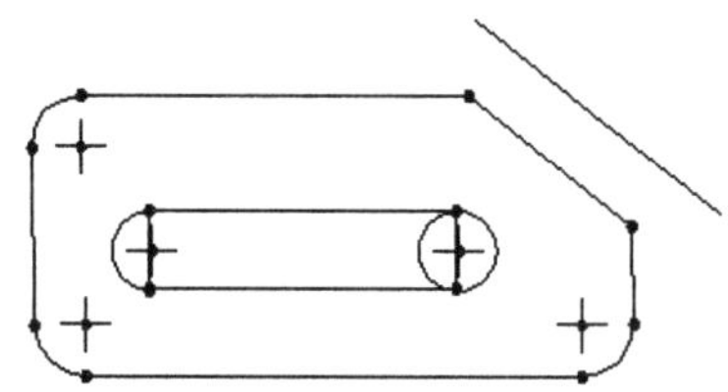

- Delete the original angled line from your sketch. Use the *Trim/Extend* option to extend the original lines on your sketch so that they intersect the offset angled line, then trim the new angled line to its intersections with the original lines. When finished, your sketch should look like the following figure:

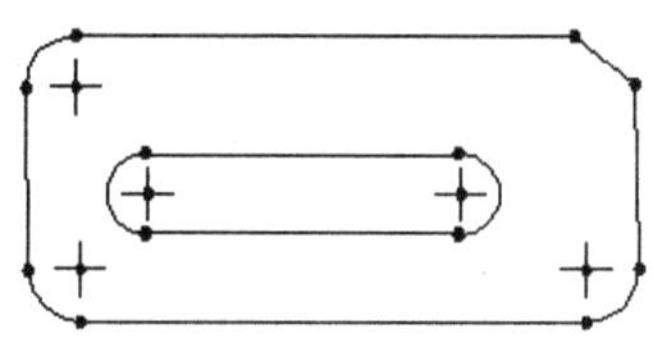

4.3.2 Extruding a 2-D Shape

Extrusion was briefly discussed at the beginning of this chapter. If you look in a dictionary, extrusion is defined as "pushing back." In this case, a 2-D shape is pushed back (or forward) a specified amount in a direction perpendicular to the plane in which the shape was sketched. As the shape is pushed back, the endpoints of the lines, arcs, and other entities that make up the shape are connected across the third dimension, forming a solid object.

Extrusion of a 2-D shape can be performed to create a new part as shown in Figure 4–1, or it can be used to modify an existing part by either protrusion or by cutting out. With protrusion, the 2-D shape is drawn on the surface of an existing object, and the shape is extruded out from that face. This results in an increased volume for the original part because material is added to it. Figure 4–9a shows a block with a sketch consisting of three circles on one face and Figure 4–9b shows the same block after the circles were extruded to form a protrusion on the object (Note that the smaller circles were not protruded the same distance as the larger circle.)

Extrusion to cut out a shape from an existing part is the opposite of the protrusion operation. With this method, a 2-D shape is sketched on the face of an existing object and is extruded into the part. As it is extruded, material is removed from the existing part in the shape of the 2-D sketch. Thus, the volume of the original object decreases. When cutting out the material, you can choose to cut all the way through the part or only partially through. Figure 4–10 shows an object with a sketch on one face and the result of extruding the indicated 2-D sketch through the part by cutting out material.

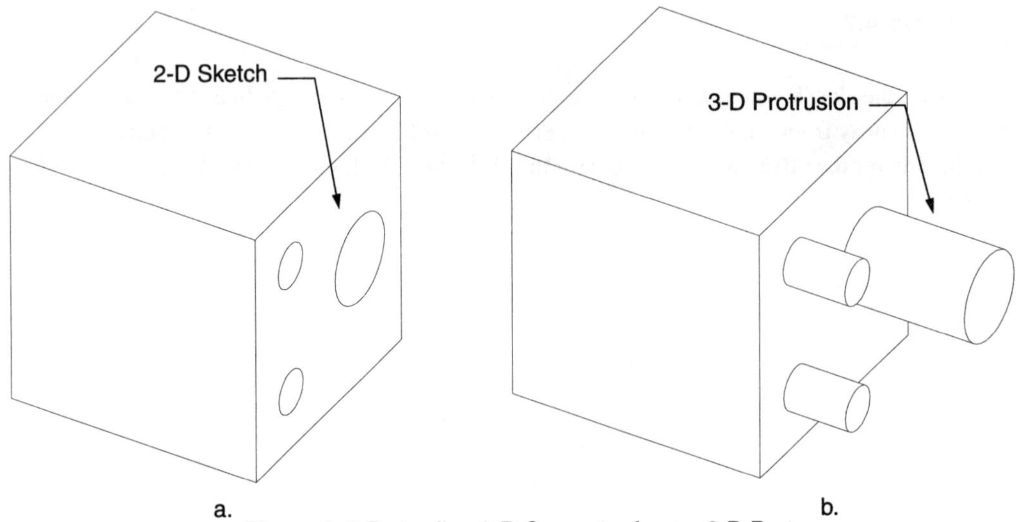

Figure 4–9 Protruding 2-D Geometry from a 3-D Part

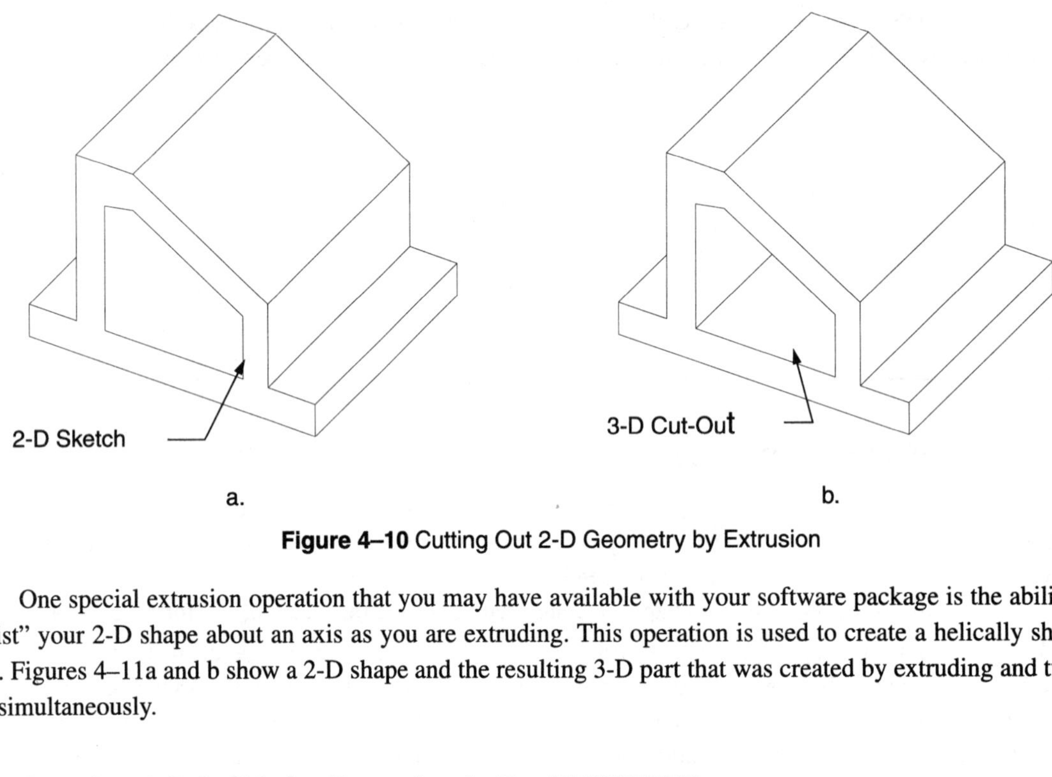

Figure 4–10 Cutting Out 2-D Geometry by Extrusion

One special extrusion operation that you may have available with your software package is the ability to "twist" your 2-D shape about an axis as you are extruding. This operation is used to create a helically shaped part. Figures 4–11a and b show a 2-D shape and the resulting 3-D part that was created by extruding and twisting simultaneously.

4.4 Creating 3-D Solids by Extrusion in Pro/ENGINEER

In order to create an object by extrusion in Pro/ENGINEER, you first sketch and dimension the section and then enter the thickness or depth of the extrusion (recall Guided Tour 2.1 where you created a simple extruded cylinder). To begin an extrusion, you must select the following commands from the Menu Manager:

Feature—Create—Solid—Protrusion—Extrude—Solid—Done

a. b.

Figure 4–11 Extruding and "Twisting" a 2-D Profile

When you use the *Extrude* command, you are asked whether you wish to create a one- or two-sided extrusion. You are then prompted to select the plane you are going to sketch the section to be extruded on (the *sketching plane*, usually *DTM3*), and the direction for the extrusion. Then, you will align the sketching plane to the screen by selecting a *reference plane* (usually by selecting *Top* and then *DTM2*). An example of a simple extrusion is shown in Figure 4–12.

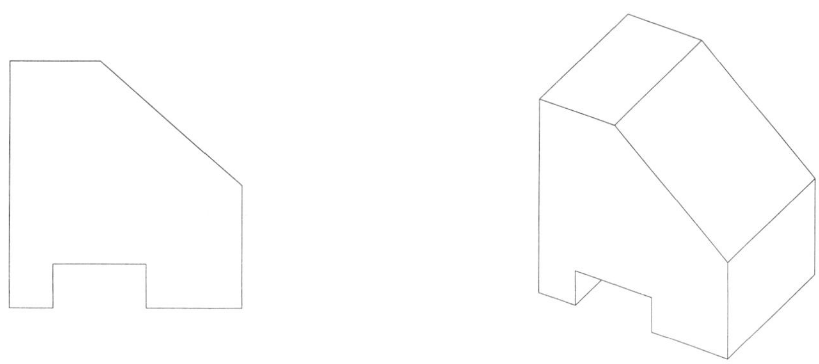

a. b.

Figure 4–12 Extruding a 2-D Shape

You can also choose to extrude two shapes at one time with Pro/ENGINEER such as an object that has an "opening" in it. To create a part with a hole in it, simply sketch the initial section with the interior hole. Pro/ENGINEER will interpret this to mean that you want the interior section to act as a "hole" in the part geometry. Figure 4–13a shows two sections that were extruded simultaneously to create the solid shown in Figure 4–13b. Note that the interior section (Section 2) created a hole in the object during the extrusion process.

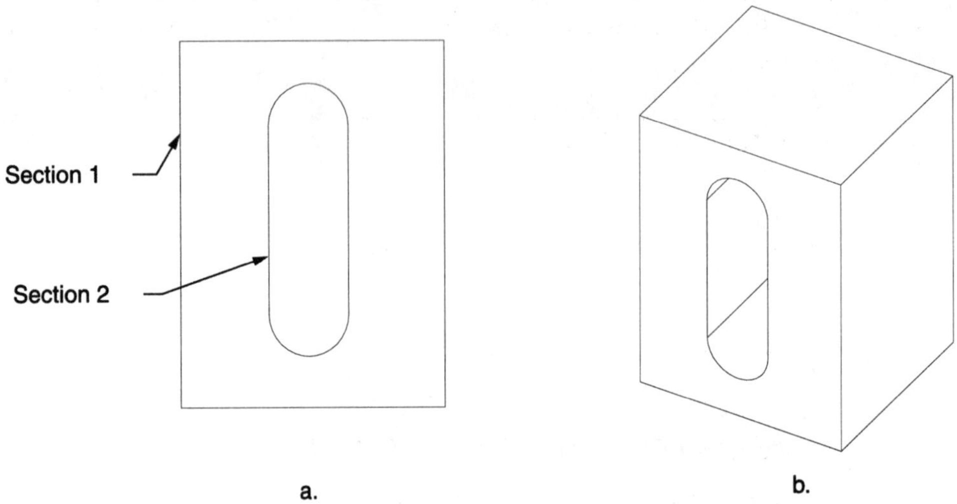

Figure 4–13 Extruding Two Sections to Form an Object with an Interior Hole

Sometimes when you create a 3-D solid by extrusion, your sketch will not have a distinct "chain" around it but will contain intersecting curves (Figure 4–14a). If this is the case, the software will not be able to figure out which of the curves you intend to include in your section for extrusion. In order to extrude this shape you must use the Geometric Tools (*Geom Tools*) to trim the unwanted line segments. Figure 4–14 shows the 2-D geometry and the resulting 3-D object that was created by extruding only the outer geometry.

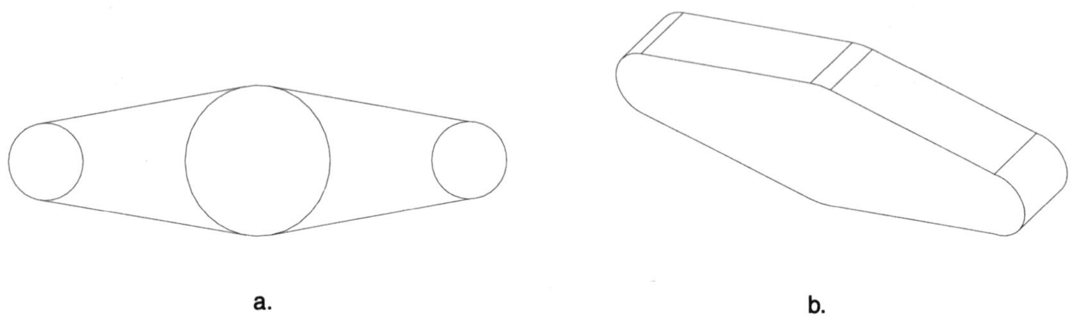

Figure 4–14

After you finish sketching, you must specify the extrusion depth in a direction perpendicular to the plane of the section. Most of the time, you will use the *Blind* or *2 Side Blind* option for extruding. Other options include *Thru Next, Thru All,* and *Thru Until. Thru Next* is used to extrude or cut the section through the next surface. *Thru All* is used to extrude or cut all the way through the object, and the *Thru Until* option is used to extrude or cut only until the next free surface is encountered.

An arrow appears on the screen perpendicular to your section to show you the length of the extrusion which will result from the parameters you specified in the form. (Use dynamic viewing, the control key and the middle mouse button to rotate the view and see the arrow length.) Once you have defined all the required parameters needed to create the extrusion, select the *Preview* button on the Feature Elements Window. If you want to make any changes to your selection, you can highlight the listed item in the window and click on the *Define* button (see Figure 4-15). This allows you to redefine any particular attribute of the extrusion. Once everything is to your satisfaction, select OK.

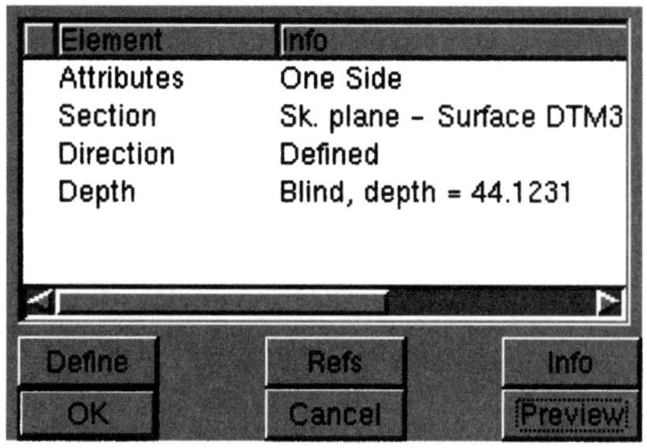

Figure 4.15 Feature Elements Window

Guided Tour 4.3

- You are now going to extrude the 2-D geometry you created in the previous Guided Tours to create a 3-D part with a slotted hole through it. If you closed your session from Guided Tour 4.2, you will need to start a new part by selecting: *Create—Solid—Protrusion—Extrude—Solid—Done*.
- After you have fully constrained your part (Intent Manager was ON, right?), select *Done* from the Menu Manager and work your way through the various windows that pop up.
- Let Pro/ENGINEER select the distance. Once the extrusion is fully defined, select *Preview* from the Feature Elements Window.

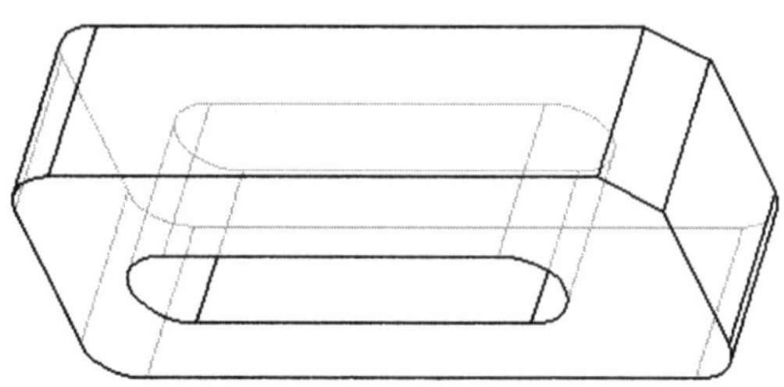

- Try redefining the various element within the Feature Elements Window.
- Press *OK* to finish the part.

4.5 Sketch Planes

When working with solid modeling software, you must remember that you are working in three-dimensional space. Sketches are typically made on a two-dimensional plane somewhere in space. Therefore, one of the first things you must do before making most sketches is define the plane on which you are sketching. Otherwise, the computer will not know exactly where your sketch is located with respect to other geometric entities on the screen or stored in memory.

Most fully 3-D software packages have a predefined working plane on which you can sketch. Typically, it is assumed that you are sketching on that plane unless you indicate otherwise. You can move this working plane around in three-space by rotating or translating it so that you are not confined to sketching from the same planar location. Sometimes you can attach the working plane to the surface of an existing object allowing you to sketch directly on that surface. Figure 4–16 shows a block with a two-dimensional shape sketched onto one of its surfaces. Note that from this view, it is impossible to tell exactly where the shape is located in three-space, however, since you know that it was drawn on the indicated surface, its location is fixed in space.

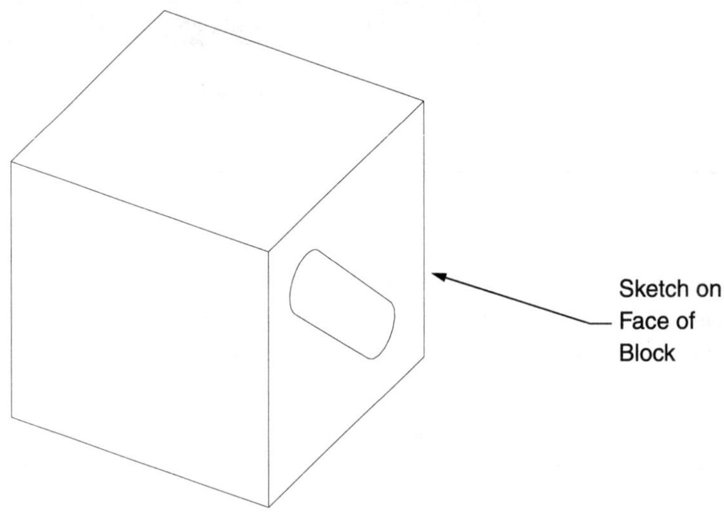

Figure 4–16 Block with 2-D Geometry Sketched on One Surface

4.5.1 Sketching on a Surface of an Object in Pro/ENGINEER

There are two common types of working planes used in Pro/ENGINEER. The first type is the set of datum planes. One of the first steps normally taken in creating new parts is the creation of datum planes. This gives the designer a working frame of reference. Sketches are made on one of the planes, then references to the others are indicated through dimensional constraints.

The second common working plane is created by selecting the surface of an existing part. This is used when you want to create a new feature on the part. To create a sketch on the surface of an existing part, you run through the standard *Feature—Create—Protrusion—Extrude—Solid—Done* sequence. When prompted for the "front" sketching plane, you can *Query Select* one of the original part's surfaces. Then, select Default as the reference, and your selected sketching plane will be aligned to the screen.

4.6 Revolving a 2-D Shape

As described earlier in this chapter, revolving a 2-D shape to form a 3-D solid is another commonly used method in part creation. This method of part creation is generally used for parts that have a basic cylindrical shape. When creating a solid by profile revolution, you typically specify the number of degrees through with you wish to revolve the profile. You must also specify the axis about which you want the profile to revolve. Figure 4–17a shows a 2-D profile and Figures 4–18b & c show the result of revolving the profile about the axis labeled "Y" by 180 degrees and 360 degrees, respectively. Figure 4–17d shows the object that results by revolving the profile 360 degrees about the axis labeled "X." Notice that each of these operations results in final parts that are very different from one another.

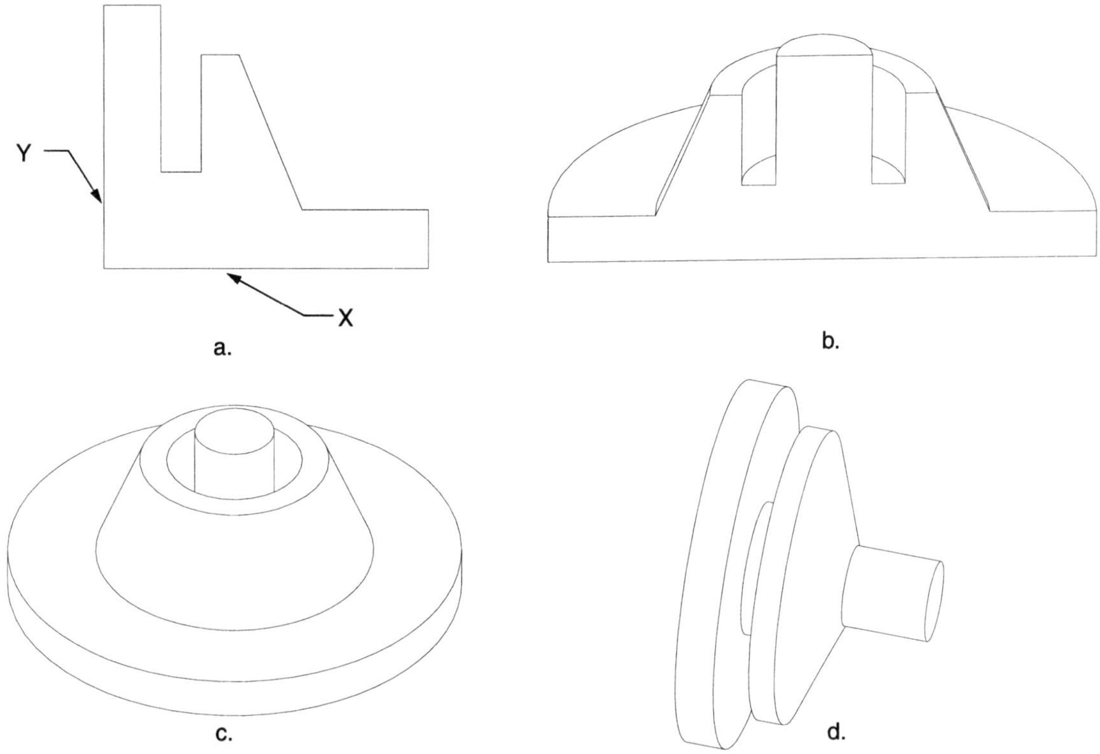

Figure 4–17 Revolution of 2-D Geometry

Cylindrical parts often have a hole through their center. This feature can be created as the 2-D shape is revolved to form the solid. If the axis of revolution is not located on the 2-D shape itself, but is located some distance, x, away from it, then a part will be created after revolution that has a hole through its center of diameter 2x. Figure 4–18 shows a 2-D shape and the 3-D part that results by revolving the shape about the labeled axis of revolution shown in the figure.

You may encounter a problem in creating objects by revolution if you attempt to create *self-intersecting* parts. This typically happens when you try to revolve a portion of the profile onto the 3-D solid that you are in the process of creating. The error occurs when the axis of revolution is on the interior of the 2-D shape you are revolving. Figure 4–19 shows a 2-D profile and an indicated axis of revolution. Note that if you attempted to revolve this profile about the indicated axis, a portion of the profile would interfere with the 3-D object geometry as the solid was swept. This is known as a self-intersecting object, and it is not permitted.

With some software packages, you are able to translate your 2-D shape along the axis of revolution as you are revolving it. This method is ideally suited to creating springs or other helically shaped solids. Figure 4–20 shows the creation of a spring by this method. In this case, the 2-D shape was revolved five complete revolutions (a total of 1800 degrees) and was simultaneously translated along the axis of revolution. The result is a "spring" that has a total of five turns along its length. When creating springs or similar objects by this method it is important that you do not try to create self-intersecting parts. If this occurs, you must either decrease the size of the profile that is being revolved, increase the translation distance, or decrease the number of revolutions in order to fix the problem.

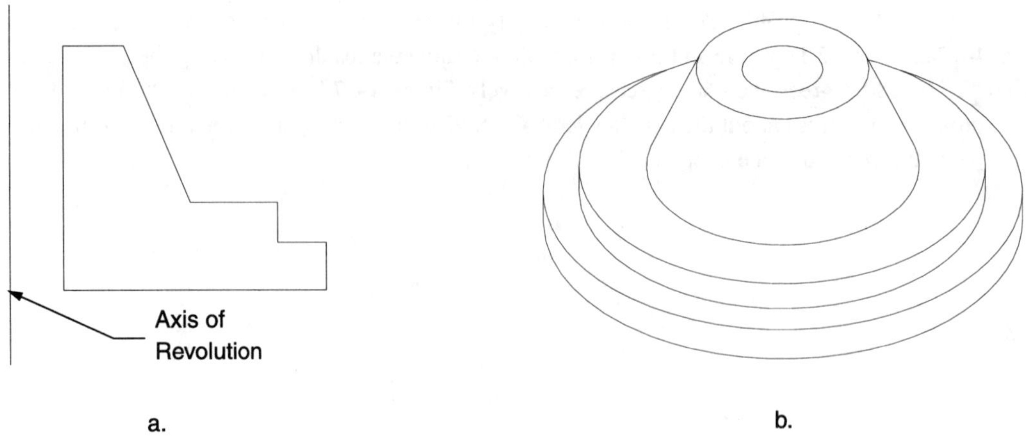

Figure 4–18 Profile Revolved about Axis not on the Profile

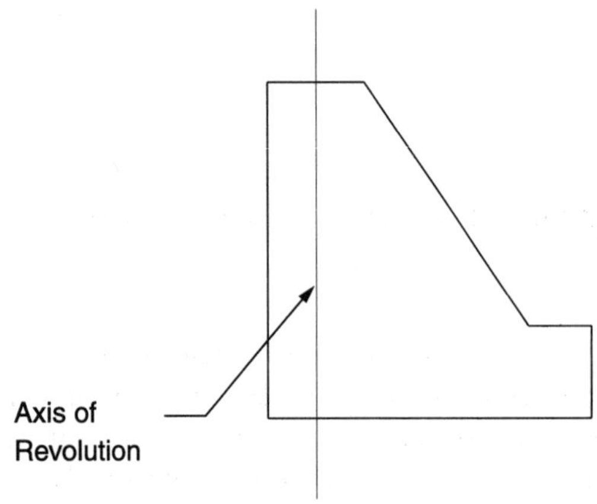

Figure 4–19 Axis of Revolution Location for Self-Intersecting Part

4.6.1 Creating 3-D Solids by Revolution in Pro/ENGINEER

The *Revolve* command in Pro/ENGINEER is found using a similar command sequence to the Extrusion. The command sequence is *Feature—Create—Solid—Protrusion—Revolve—Solid—Done*. If you are using the Intent Manager and the datum planes (as you should), you will first need to select horizontal and vertical references. It is most convenient to select the vertices created by the datum planes.

Once the references for the Intent Manager are defined, the next thing you need to do is sketch your center line. From the Menu Manager, select *Sketch—Line—Center Line.*

This is perhaps the most overlooked step in creating a revolved shape. Luckily, Pro/ENGINEER will catch the missing center line before you try to revolve your part. Once the center line has been sketched and constrained, you simply sketch your section to extrude, select *Done*, indicate the angular rotation (90°, 180°, 270°, 360°, etc.), and select *Preview* and *OK* from the Feature Elements Window.

The creation of a helical "spring"-shaped part requires the use of *Advance—Protrusion*. The command sequence used to generate a simple spring is: *Feature—Create—Solid—Protrusion—Advance—Solid—Done*. The Menu Manger changes we will then need to select are: *Helical Sweep—Constant—Thru Axis—Right Handed—Done*. Any of these options can be changed, and you should test the other options, but for the sake of discussion, we are limiting ourselves to the previous command sequence.

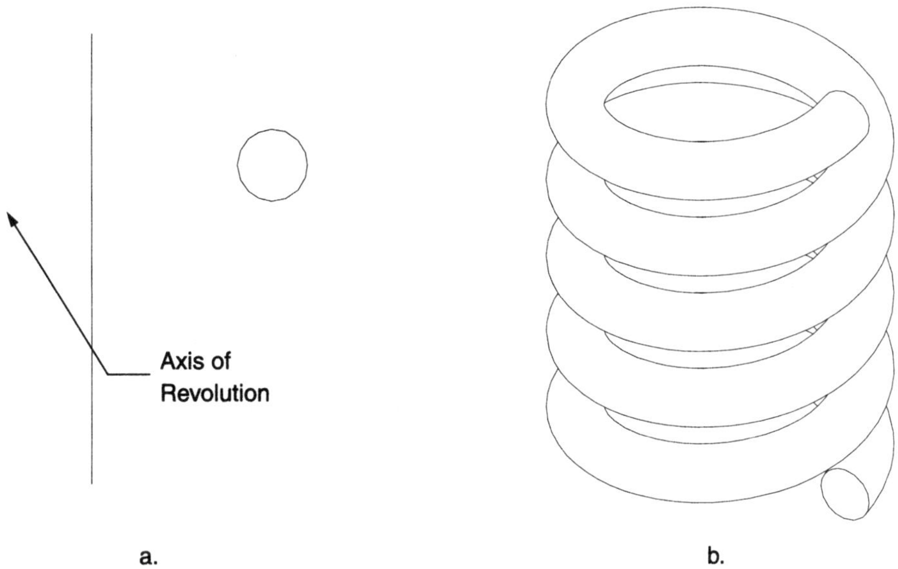

Figure 4–20 Translation of a 2-D Profile During Revolution

There are three things that Pro/ENGINEER requires you to define: the Sweep Profile, Pitch, and Section. The Sweep Profile consists of defining the center line and the outer contour of the coil. The Pitch represents the translational change for each revolution. The Section simply refers to the shape that is to be protruded around the helical path. Figure 4-21 shows the relationship between Sweep Profile and the final part.

Once you have created revolutions, you are encouraged to experiment and find out what types of limitations the various components of revolved sweeps have. Can the Sweep Profile loop back on itself? What happens if the Section diameter is larger than the Pitch? Try it and find out.

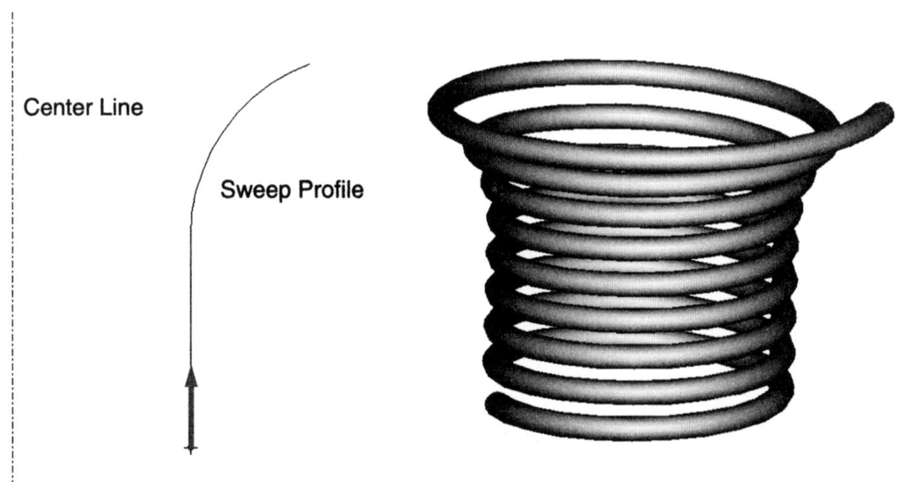

Figure 4-21 The Relationship Between the Sweep Profile and the Final Part

Guided Tour 4.4

- Create a new part using: *Feature—Create—Solid—Protrusion—Revolve—Solid—Done.*
- If you are using datum planes and Intent Manager, define the horizontal and vertical references.
- Sketch a center line by selecting *Sketch—Line—Center Line* from the Menu Manager.
- Create the section profile as follows:

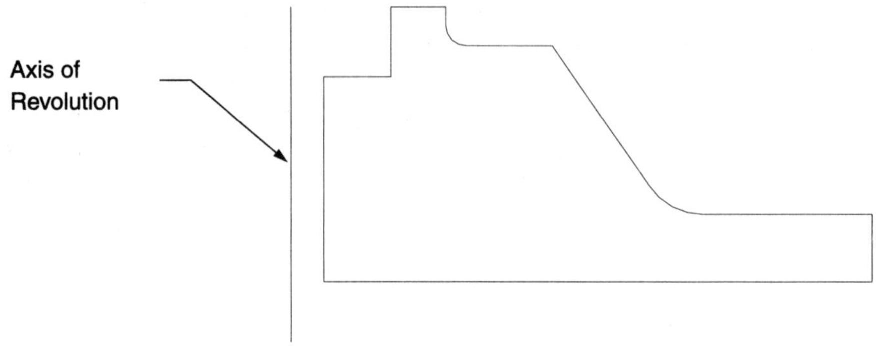

- Revolve the section geometry 360 degrees about the indicated axis of revolution to obtain the 3-D geometry shown in isometric below.

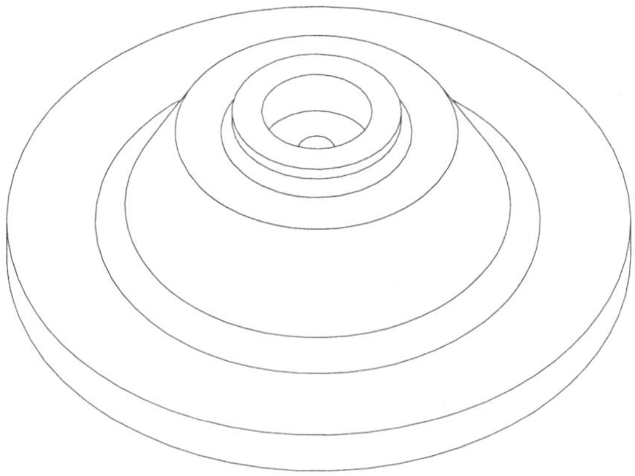

- Now cut holes in the object. Specify *Thru All* as your Distance option in the Extrude Section form. When you are finished, your object should look similar to the one shown below:

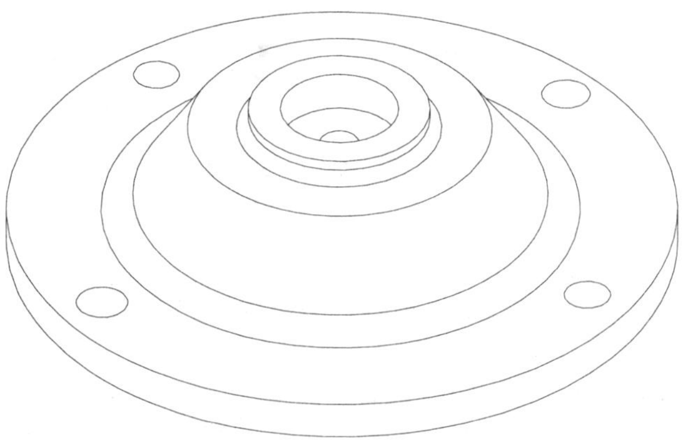

- Experiment with some of the viewing options in Pro/ENGINEER with this geometry. Display a shaded image of the object to verify for yourself that the holes were properly cut through the original object. Obtain a plot to submit to your instructor. Exit Pro/ENGINEER.

4.7 Modeling Strategy

The methods you choose to create 3-D solids will depend on the shape of the geometry you are trying to model. Many times, there will be more than one viable method you can use to achieve the desired result. You must choose the method with which you are the most comfortable. Figure 4–22 shows two methods for creating a simple cylinder. By the first method, a circle was drawn and extruded to form a 3-D solid. This method is especially useful when you are creating cylindrical holes in an existing object by sketching on a planar surface and extruding to cut out. With the second method, a rectangle was drawn and revolved about one of its edges to form a 3-D object. Both of these methods resulted in the same object, however, different procedures were followed by each.

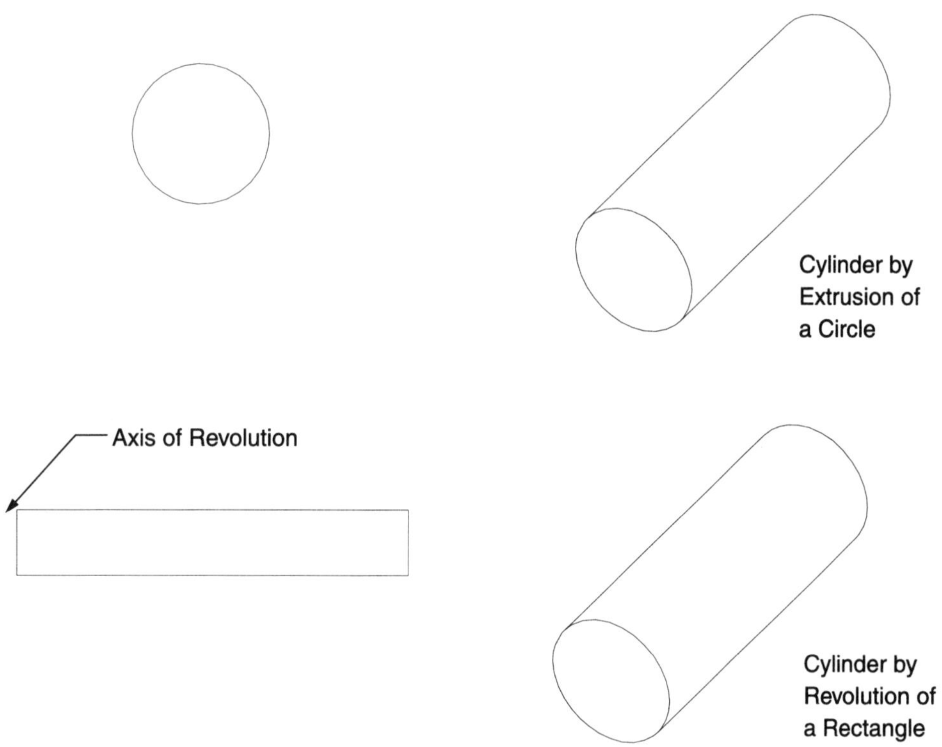

Figure 4–22 Creating a Cylinder by Extrusion and by Revolution

When you start creating a model, it is important that you think and plan before you sit down in front of the computer to begin your work. Often there is more than one viable method for creating a solid object, and you must plan your strategy. If you were to actually build something like a table, you would first plan how you were going to go about it, purchase all of the materials ahead of time, and probably make some sketches of how you were going to construct the table. Likewise, you would look at the object you are trying to build, make some sketches of how you are going to go about the modeling process and *then* actually sit down in front of the computer to "build" it. In this section, you will learn about different ways that you can create some objects. Basically, you can usually create a model by starting with a larger object and removing volume from it, or you can start by building up or adding to your initial shape. Sometimes it is easiest to use a combination of these strategies. The method you choose will depend on your personal preferences among other things. As each method is presented in the following examples, you may wish to work along at your computer workstation to become more familiar with the solid modeling process.

Example 1

How would you go about modeling the object shown in Figure 4–23? Two methods for creating this object are outlined in the following.

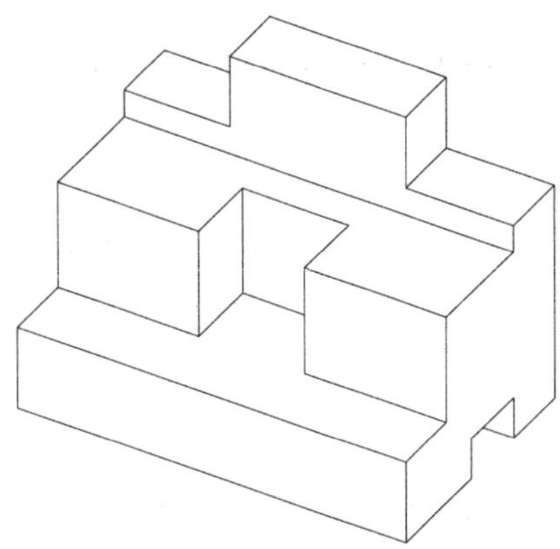

Figure 4–23 Object for Example 1

Method 1

- Start by sketching the shape of the object from the side and extruding it to form a 3-D solid:

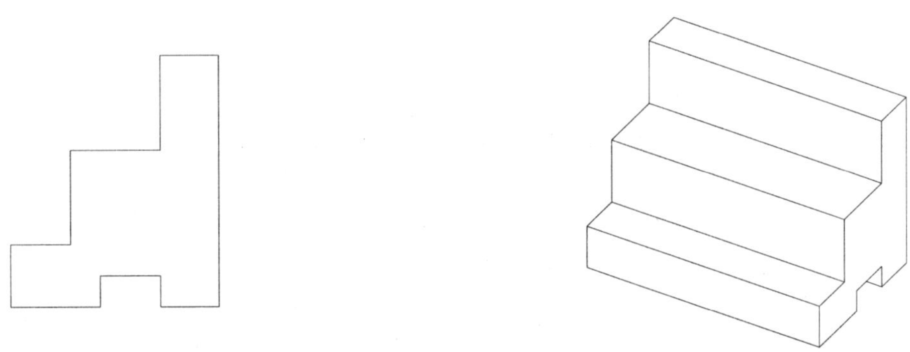

- Sketch two rectangles on the vertical face of the back portion of the object you just created, and extrude them to cut out the object:

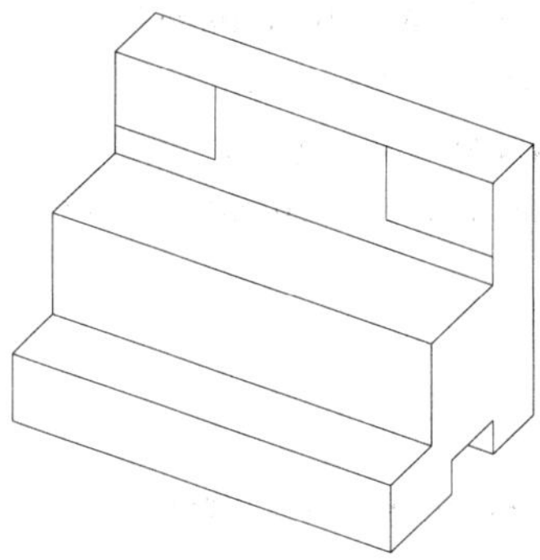

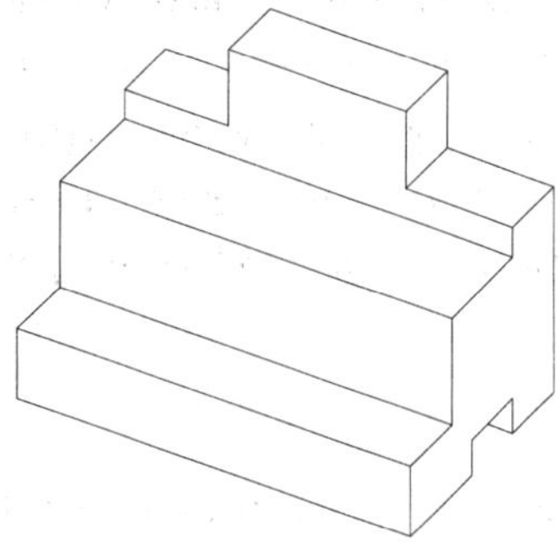

- Sketch a rectangle on the horizontal surface of the middle portion of the object, and extrude it to cut out from the object:

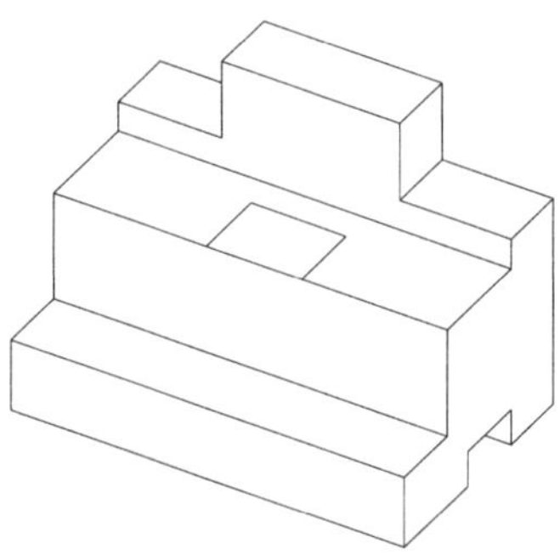

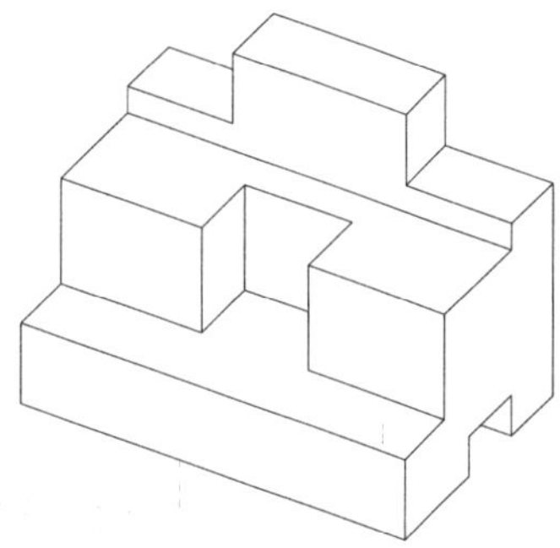

Method 2

- Sketch the shape of the object as seen from the front, and extrude it to form a solid:

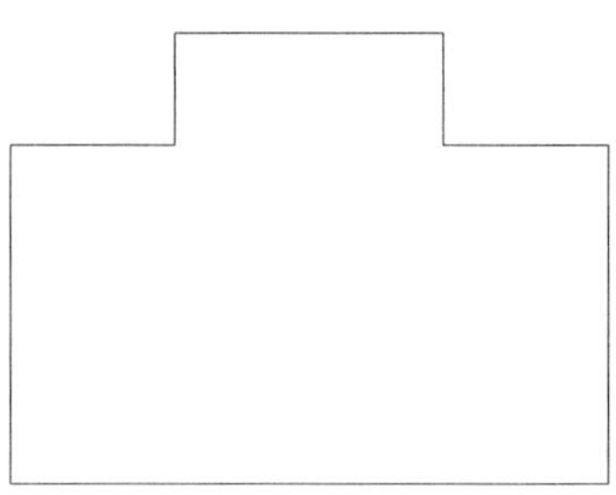

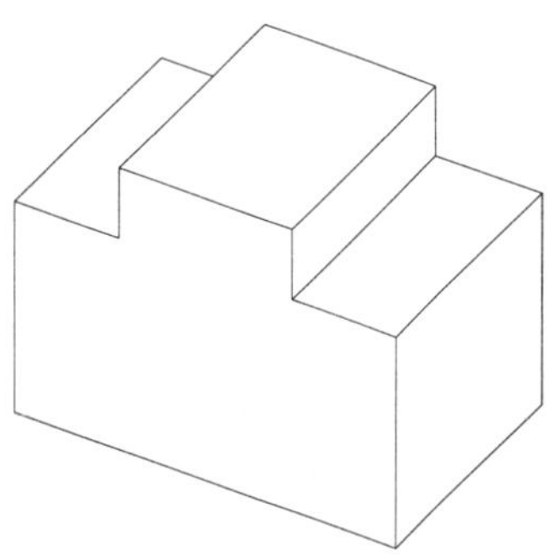

- Sketch a rectangle on the side face of this 3-D object (Note that the rectangle extends beyond the limits of the surface itself), and extrude it to cut out from the object:

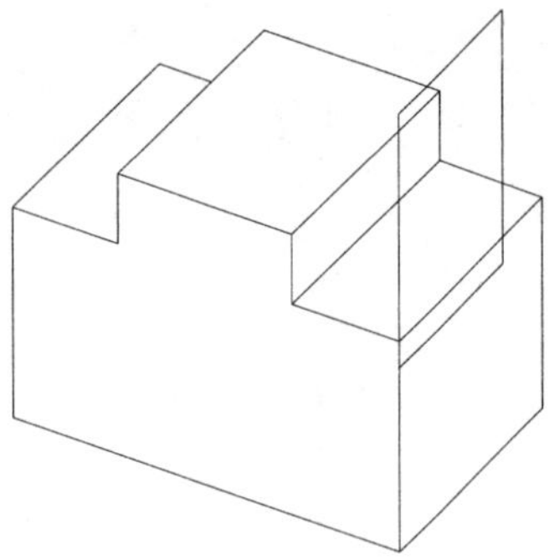

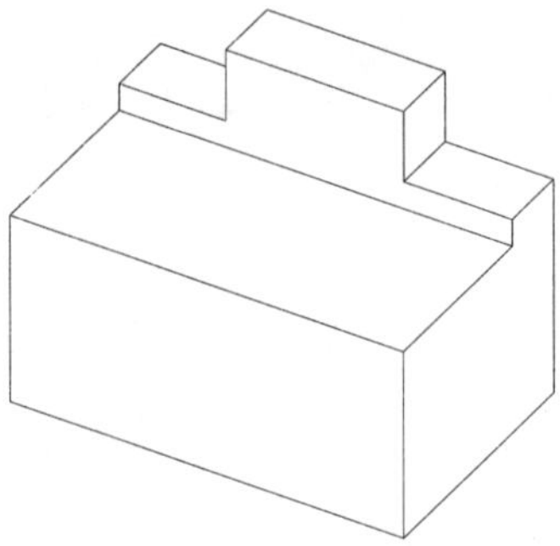

- Sketch a T-shape on the horizontal surface of the front portion of the object, and extrude this shape to cut out the object:

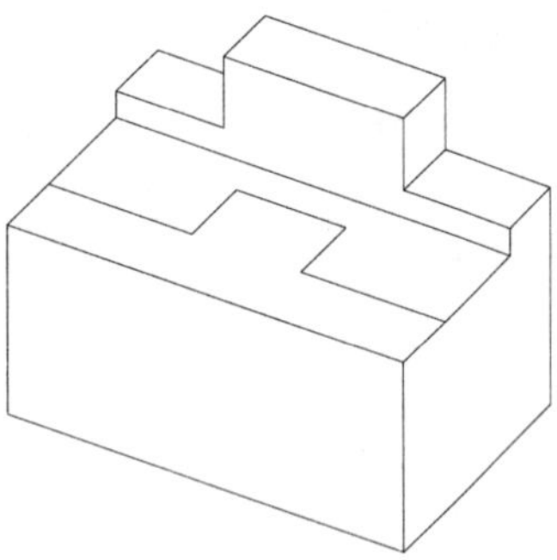

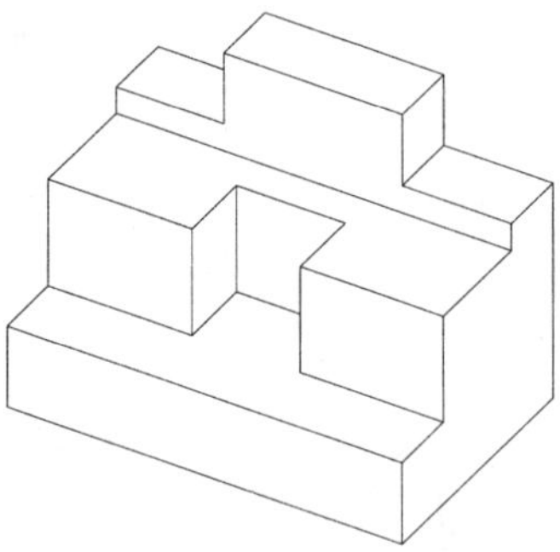

- Sketch a rectangle on the side surface of the object, and extrude it by cutting all the way through the object:

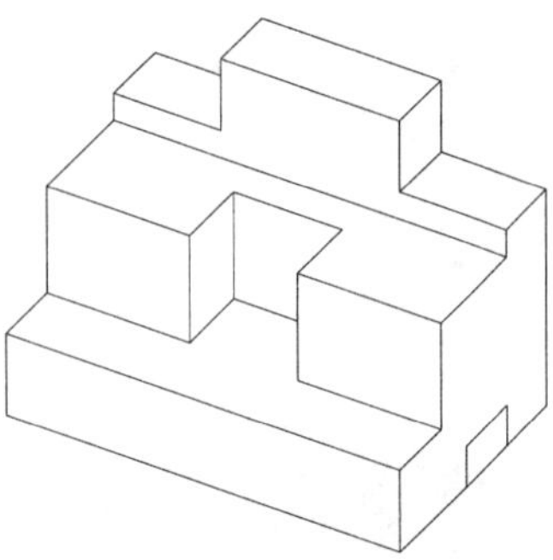

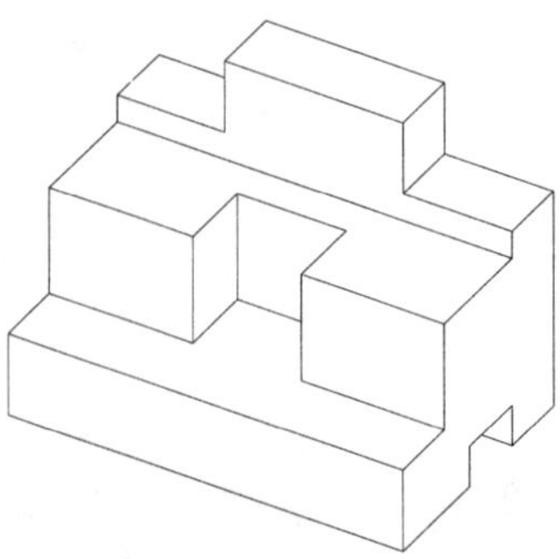

Example 2

How would you go about creating the object shown in Figure 4–24?

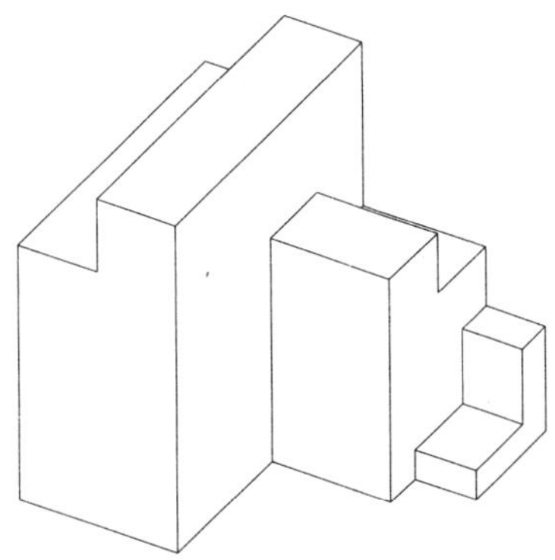

Figure 4–24 Object for Example 2

- Start by sketching the shape of the left portion of the object and extruding it to form a solid:

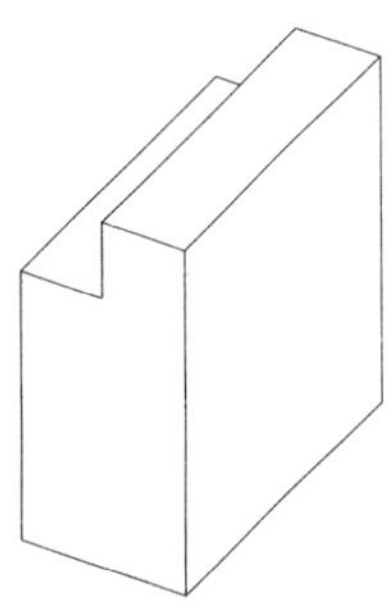

- Sketch on the vertical face of the side surface, and extrude it to protrude from the original object:

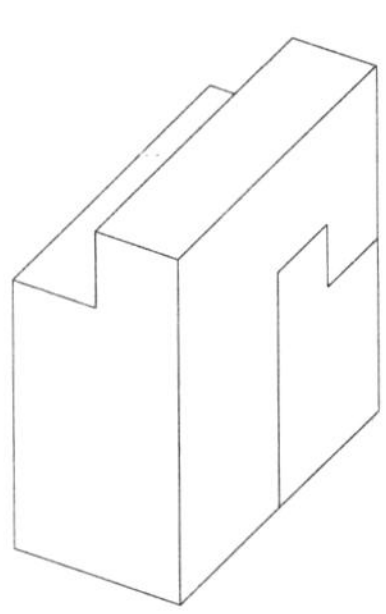

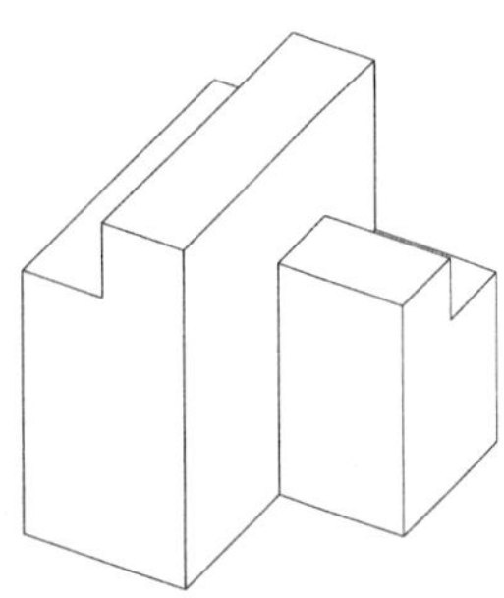

- Sketch the third shape on the new vertical face of the side surface, and extrude it to protrude from the object:

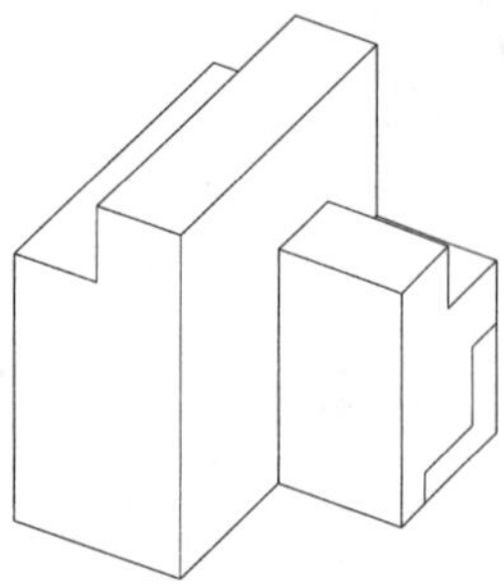

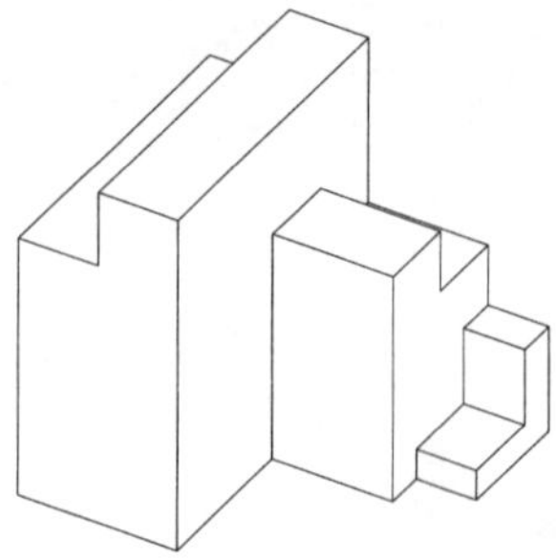

Example 3

As a final example, how would you create the object shown in Figure 4–25?

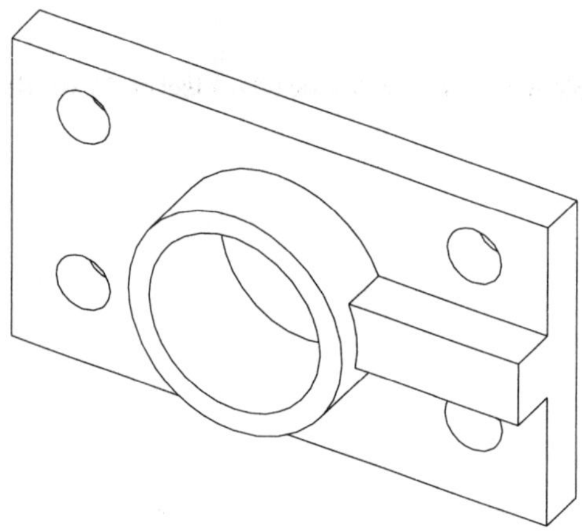

Figure 4–25 Object for Example 3

- Start by sketching a rectangle and four circles and extruding this to form the base plate with four circular holes (Note that you could do this operation in two steps—extrude a rectangle, sketch four circles on its face, and cut them out, if you wish):

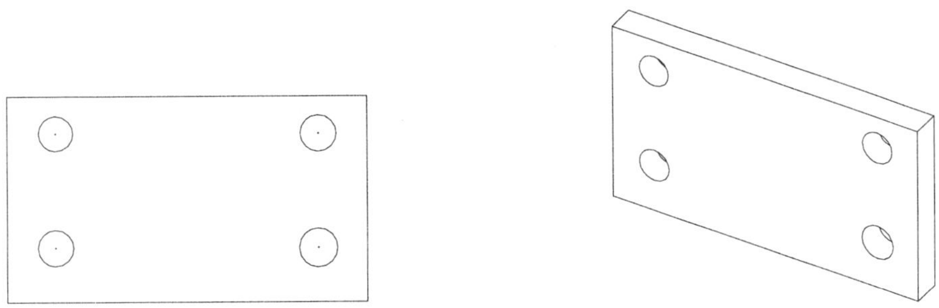

- Sketch a large circle in the center of this plate and protrude it upwards from the base:

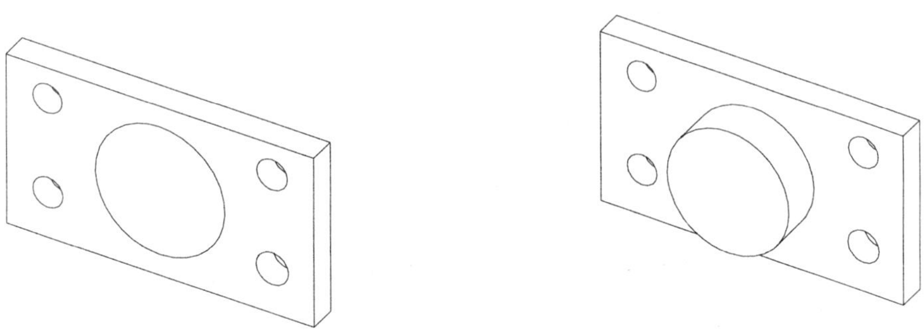

- Sketch a rectangle on the face of the base and protrude this up to form that feature on the object. Have the rectangle extend "inside" the boundary of the cylindrical feature (Note that the wireframe view of the object with the rectangle sketched on it is shown for clarity):

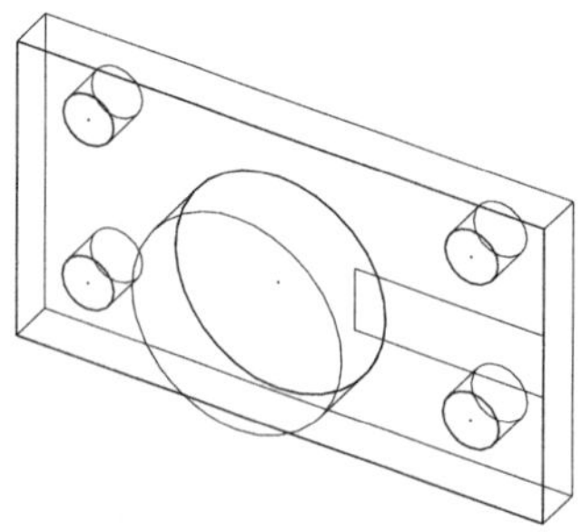

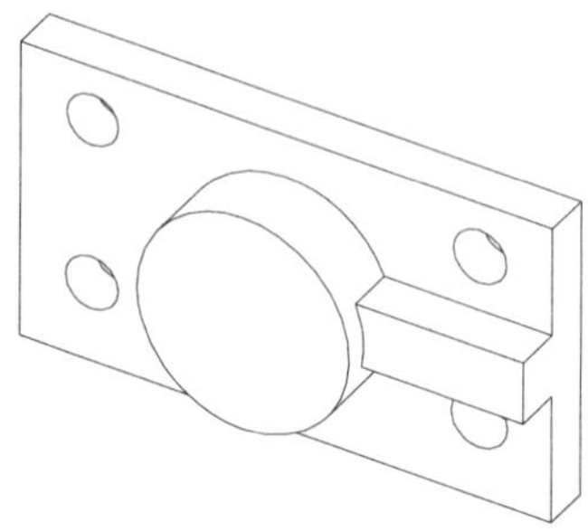

- Sketch a circle on the end of the cylindrical feature, and cut this all the way through the base object:

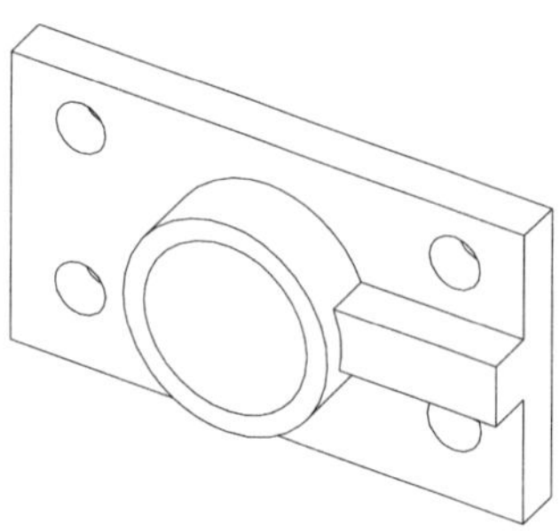

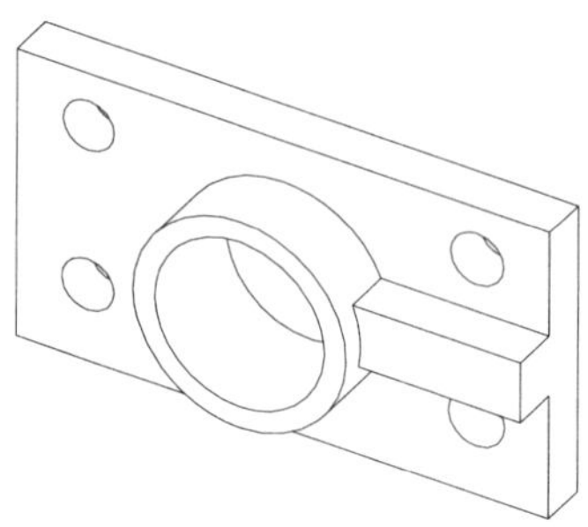

4.8 Troubleshooting/FAQs

You might encounter a problem when creating 3-D objects by extrusion or revolution if you are working with 2-D geometry that is "open." Open geometry means that there is a gap in the outline of the shape you are working with. Sometimes this gap will be invisible to your eye, even if you zoom the area of the sketch where it occurs. It is important that you start and stop subsequent curves on your shape at exactly the same point in space. The Intent Manager will automatically do this for you by "picking" the endpoints of the curves for you as you draw. Sometimes, however, depending on how you created the geometry, you may have to work to ensure that this is the case. The easiest way to accomplish this is to use the *Alignment* command found in the Menu Manager during the Sketch mode.

Problem	Potential Solution
You are trying to trim circles and you keep getting the wrong portion of the arc to remain.	You may find it easier to first *divide* the circle.
You want to create Polylines that don't start and stop at the same origin, but you can't figure out how to get out of creating the polylines.	Click the middle mouse button to signify that you are finished creating polylines.
You can't seem to pick the plane for sketching that you want.	1) Use *Query Select* and toggle through the choices, or 2) use the dynamic viewing capabilities to obtain a view of the surface that makes it easier for selection.

4.9 Chapter Summary

In this chapter you first learned how to sketch and modify 2-D shapes in Pro/ENGINEER. You then learned how to create 3-Dimensional objects from these 2-D shapes by either Extrusion or Revolution. You also learned about sketching on the plane of an existing object. Key Words for the chapter are tabulated in the following:

2-D Fillet	Points
Arc	Polygons
Axis of Revolution	Polylines
Circle	Protrude
Cut Out	Revolve
Delete	Rounds
Exterior Section	Section
Extend	Self-Intersecting Part
Extrude	Sketching
Fillets	Splines
Flip Side	Thru All
Interior Section	Trim/Extend
Lines	Trimming Lines
Offset	Trimming Circles

Exercises

On a clean sheet of paper, develop a **Plan of Action** for the object assigned by your instructor describing the steps you would follow to create the object in Pro/ENGINEER. Use sketches as appropriate in your Plan of Action. Then create the object in Pro/ENGINEER. Do not worry about the exact sizes of the features on the object, but try to make them look proportionally correct. Obtain a view of the object specified by your instructor and turn in hard copy as appropriate.

1.

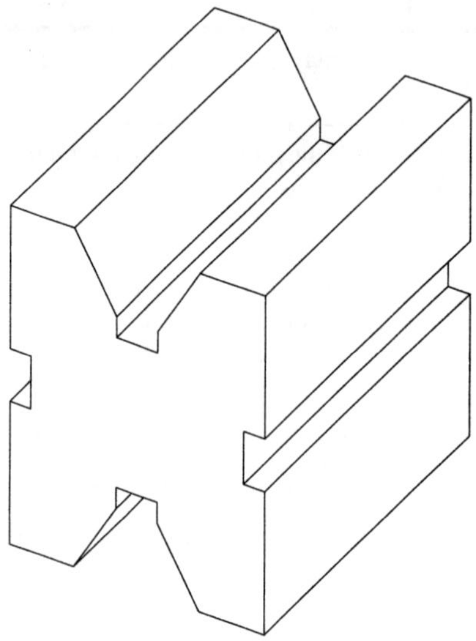

2.

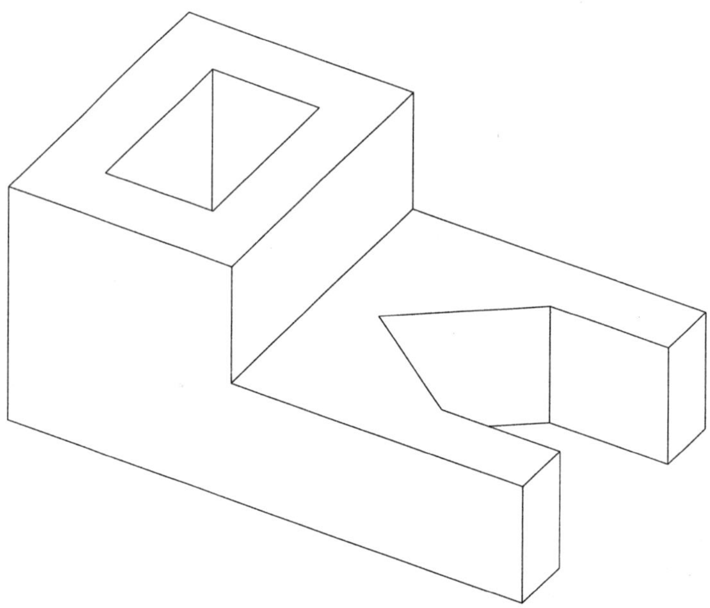

3.

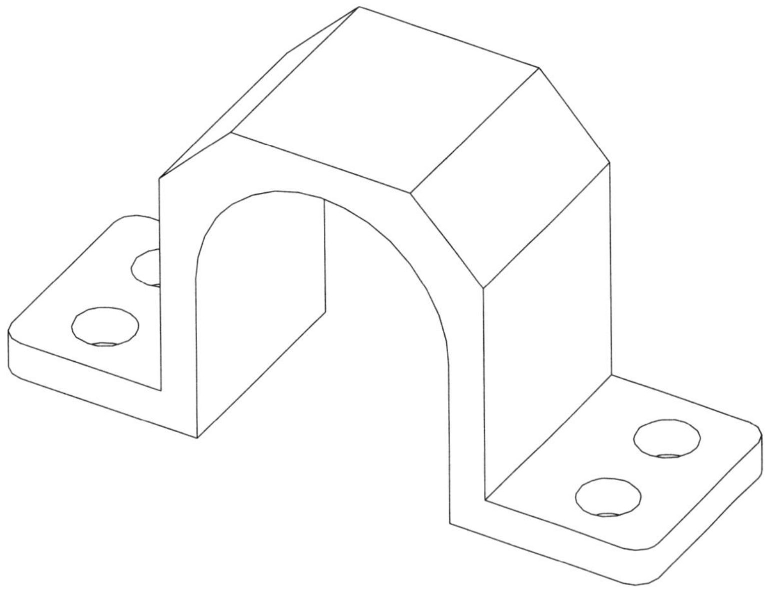

4.

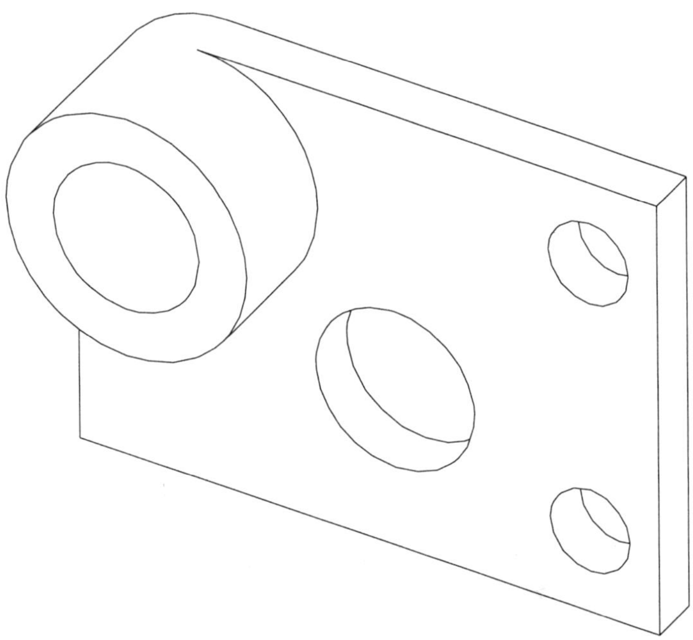

5.

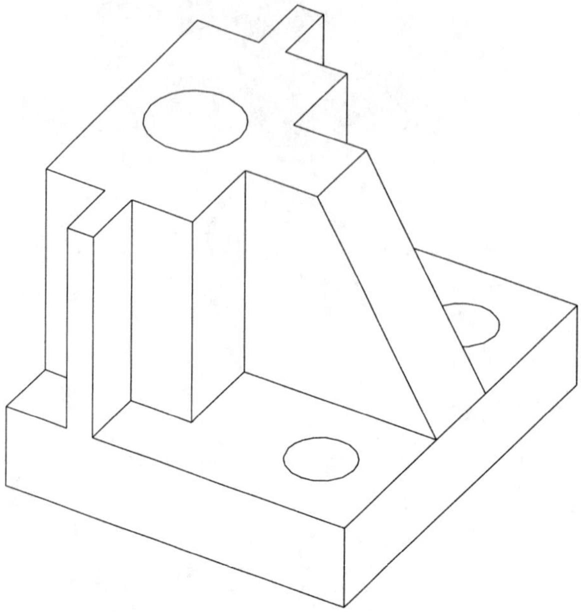

6.

This object is shown with a portion removed for clarity. Your final object should be revolved through 360 degrees.

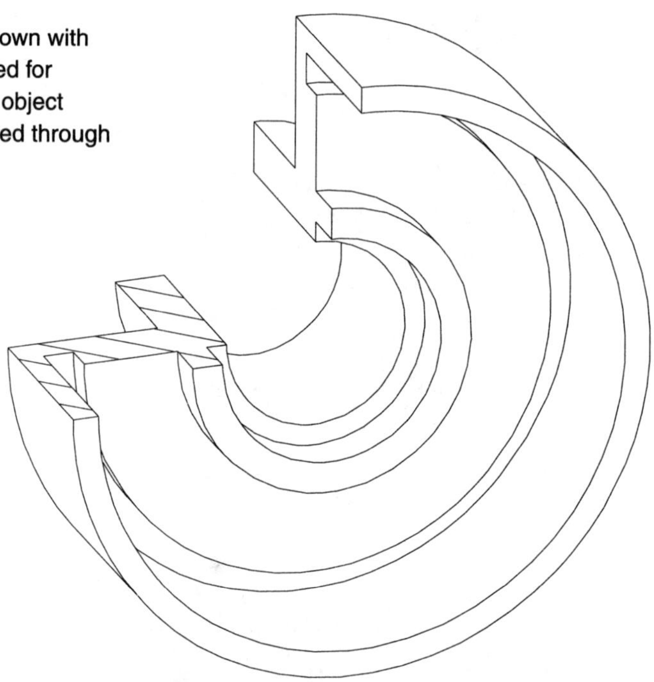

7.

This object is shown with a portion removed for clarity. Your final object should be revolved through 360 degrees.

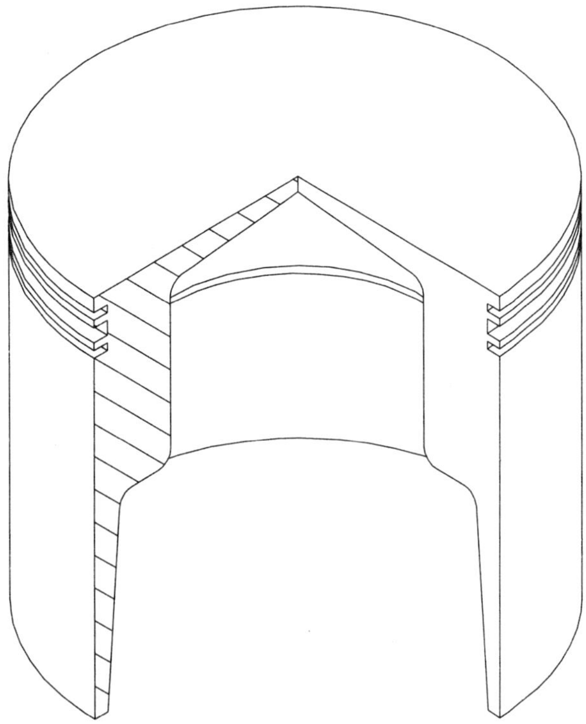

CHAPTER 5

Displaying 3-D Objects

One of the many advantages of 3-D computer modeling is the ability to display your object from virtually any angle and to control the appearance of the object by one of several techniques. In this way, you can create "snapshots" of your object to accurately portray your design. With traditional board drawing or even with computer-aided drafting, you would be required to spend a considerable amount of time creating each snapshot, line by line. Once you have created the solid object with 3-D modeling, however, you can create different snapshots by merely executing a few software commands. Thus, it is much easier to change the object display to create snapshots of the object for documentation purposes and to change the object display to assist you in the actual creation of the object.

When you create a snapshot of an object, you are concerned primarily with two settings: observer location and object appearance. Each of these will be described in separate sections of this chapter. The one thing you must remember in this discussion of object display is that the object always remains fixed in space—it may appear to move, but it does not. This chapter will discuss the primary methods for displaying your object.

5.1 Observer Location

Your location relative to the object's orientation in space will alter which parts of it you see. If you crawled underneath your automobile, you would have an entirely different view of it than if you looked from above. Likewise, if you were standing right next to the car, it would appear relatively large, but if you went to a point a quarter mile away and looked back at it, it would appear relatively small. Clearly, the car itself hasn't changed, but your viewpoint has changed so it *appears* differently to you.

Observer location is defined by the view angle (such as above, below, in front of, and so on) and the view distance or zoom level (such as near or far). Modeling software typically has several predefined view angles available as well as the ability to customize the view angle. The standard view angles include any one of the three principle orthographic views (top, front and right side) as well as an isometric or trimetric view. More preset views may be available depending on the specific modeling software. Figure 5–1 shows the same object from several different view angles. Remember that the object has not moved in space to create these different views, but you have "moved" your location in space relative to the object's orientation.

The view angle is sometimes defined relative to the 3-D coordinate axes in space. A top view is usually defined as the view obtained when your eye is located along the y-axis and you are looking down on the object from that vantage point. Another way of describing a top view is from a point in space defined as (0,1,0)—the x-coordinate of the viewer location is 0, the y location is 1 and the z location is 0. Likewise, a front view could be described as (1,0,0), a bottom view (0,-1,0), and a side view (0,0,1). Using this notation, an isometric view would be (1,1,1).

This ability to change the view angle for the object allows you to display more than one view of an object at the same time. You could choose to show the top, front, right side, and isometric views (similar to how they

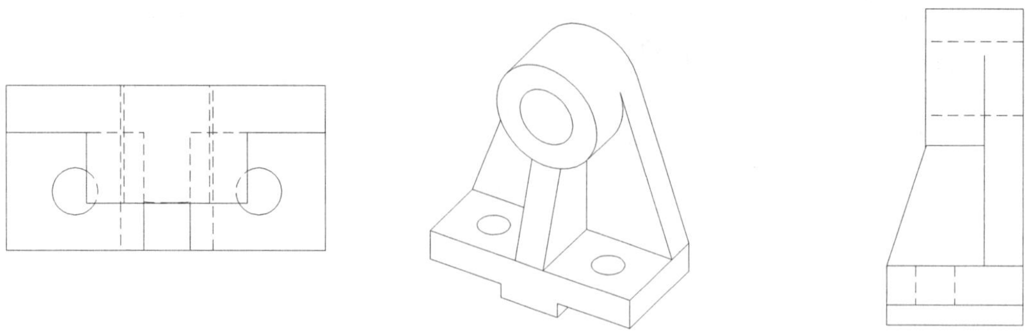

Top View Isometric View Side View

Figure 5–1 Object Observed from Different View Angles

will appear in a final drawing) as you are working on your drawing. Figure 5–2 shows four views of the object displayed simultaneously.

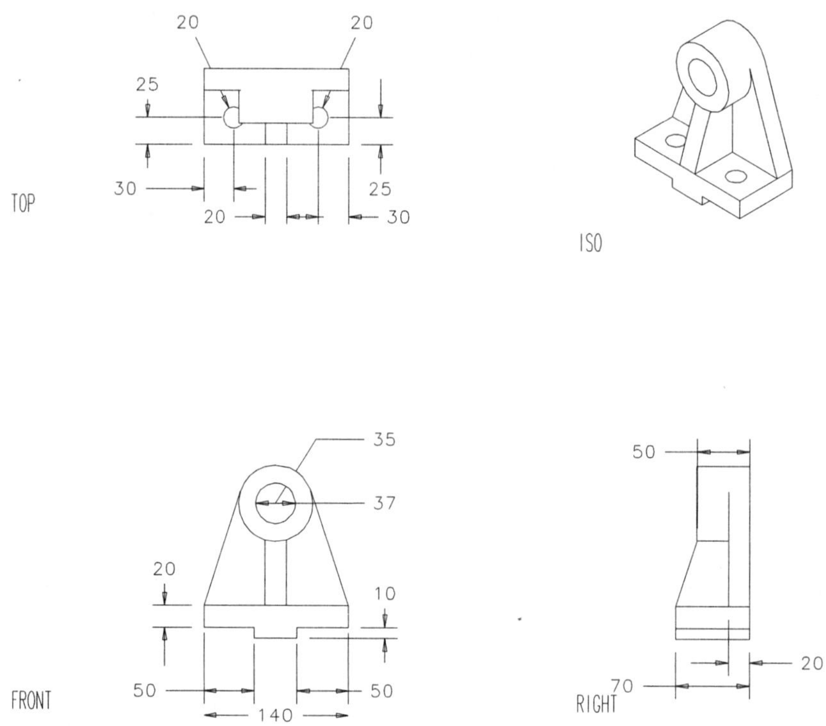

Figure 5–2 Four Views of Given Object

How large or how small an object appears on the screen is also a function of observer location. For CAD software, this is known as the level of zoom. If you have zoomed in, you are located closer to the object, and it will appear larger to you. Similarly, if you have zoomed out, you are located farther away from the object and it will appear relatively small to you. Most of the time, you will zoom in on portions of the display by "windowing" around them. One of the terms commonly used in zoom operations is "Zoom All." (This is sometimes known as *Autoscale* or *Refit*.) With this option, the object or objects you are working on will be made as large as possible to maximize the screen space. This is especially helpful if you accidentally move an object to a point in space that is located off the screen. Figure 5–3 shows the same object with two different zoom settings used for display purposes.

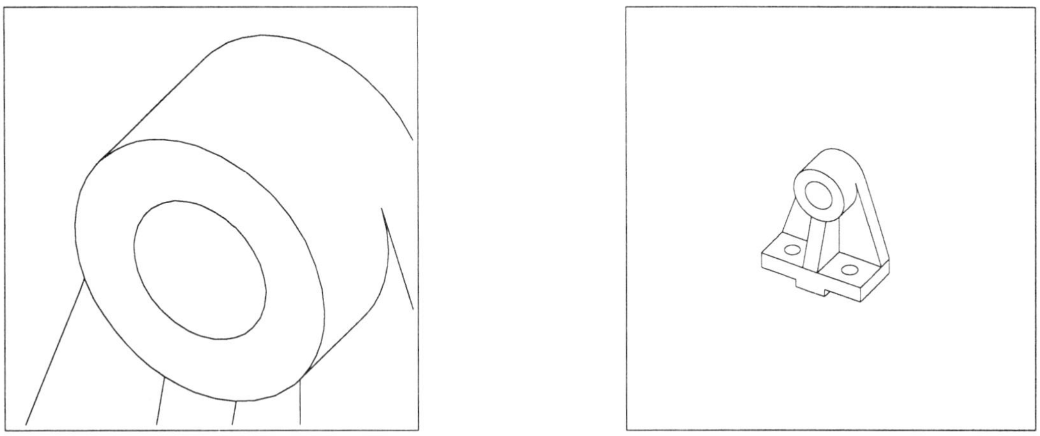

Zoom In Zoom Out

Figure 5–3 View Distance or Zoom Level

5.1.1 Setting the Observer Location in Pro/ENGINEER

Most of the options that control the display of objects on the screen in Pro/ENGINEER are found on the shortcut tool bar. These options are common to all the modes within Pro/ENGINEER so that once you become familiar with their use, you are able to use these options as you migrate from mode to mode through the software. The buttons that you use to control the observer location relative your objects are described in the following:

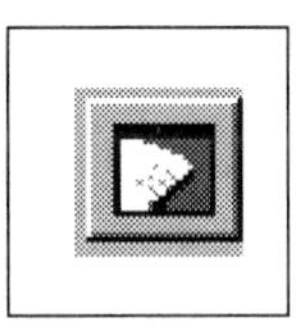

Redisplay—This button redisplays the screen. It is useful for removing clutter that has appeared on the screen during object creation and editing. The activation of this button also turns off the zoom commands.

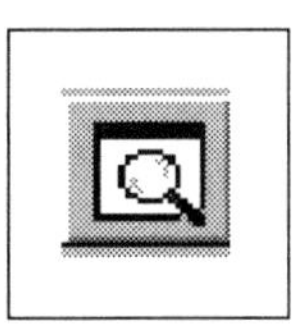

Zoom All—This command zooms the display to show all of the current objects (or all of a single object) and completely fill the graphics region. This option will become useful as your designs grow in complexity.

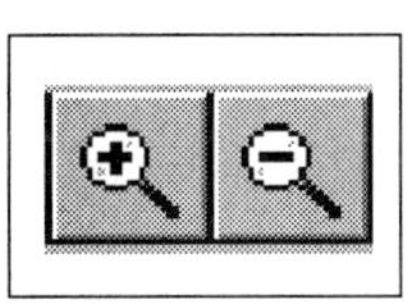

Zoom In/Out—When *Zoom In* (+) is selected, you will be prompted to use the mouse to drag across the area you wish to zoom into. The area within the zoom window will then be enlarged to fill the graphics region. To reduce the zoom by half, press the *Zoom Out* (–) button. This doubles the distance from the observer to the object.

Orient Model—Select this button to align your model such that one of its sides is parallel to the screen. Upon pressing this button, you will see a dialog box prompting you to select the front surface (the surface to be made parallel to the screen). You will then be asked to select another surface, normal to the first surface. Test this command out for yourself.

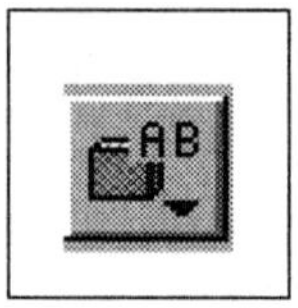

Default View—This button is used to select previously saved views. If you have not created any views, the only available option is *Default View.* This displays your object in trimetric view. To set the default view to isometric, you must change the setting found in the pulldown menu *Utilities—Environment—Default Orientation.*

Guided Tour 5.1

- In order to experiment with the display options found in Pro/ENGINEER, you will need to create a 3-D object on the screen.
- Create a cylinder with a diameter of 120 mm and a height of 360 mm.
- Experiment with the buttons that control the observer location including: *Zoom, Zoom All, Redisplay,* and *Orient Model.*
- Change the *Default View* setting from trimetric to isometric.
- While holding the *Control* key down, press the left mouse button and slide the mouse across the screen. What happens? Try the same with the middle and left mouse buttons.

5.2 Object Edge Appearance

There are many different methods for displaying solid objects on the computer screen, but the primary differences involve how the surfaces and edges defining the object appear to the observer. The edges of an object will usually appear as lines or curves, but the differences are usually apparent in how the *hidden edges* appear on the screen. Hidden edges are defined as those edges that are "behind" the faces of the object relative to the current setting for the observer view angle. The most common choices for the hidden line display are to show them as solid lines, as dashed or grey lines, or as invisible lines. When hidden lines are shown as solid lines, this is sometimes referred to as the *wireframe view* of the object. In a wireframe view, the object appears to be hollow and made up of thick wires in a framework along its edges (like a box kite). Figure 5–4 shows an object with the hidden edges displayed as solid, dashed, and invisible lines.

Another difference in edge display for computer models occurs when you choose to show it with or without perspective. Perspective display shows all of the lines on the object receding to the same point on the horizon. Figure 5–5 shows an isometric view of an object with perspective on and with perspective off. Note that these two different displays of the object are similar but appear slightly distorted relative to one another.

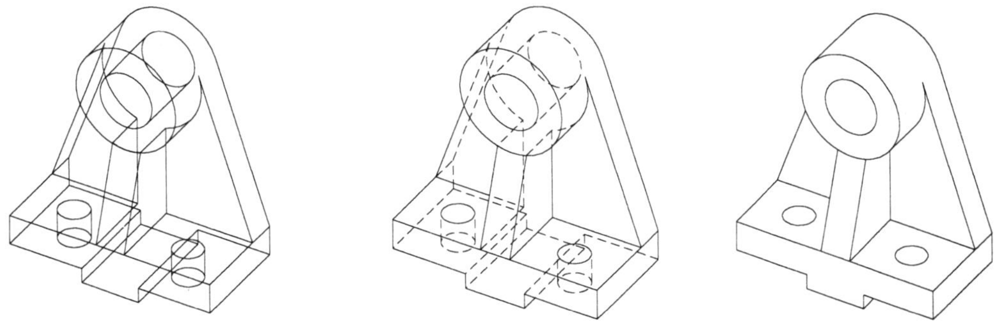

Hidden Edges Solid (Wireframe Display) | Hidden Edges Dashed | Hidden Edges Invisible

Figure 5–4 Various Hidden Edge Appearances

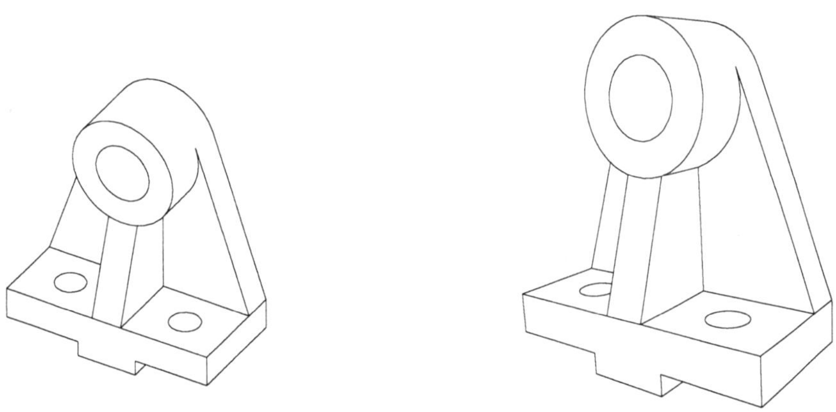

Perspective Off | Perspective On

Figure 5–5 Perspective Display

In Pro/ENGINEER, displaying an object with "Perspective On" requires the use of an advanced view option. To view an object in perspective, go to the pulldown menu *View* and select *Advanced—Perspective.* A new dialog box should appear. Here you can set your *Eye Distance, Zoom, View Angle*, etc. The dynamic viewing controls (control key and mouse buttons) work with perspective on. The *Eye Distance* and *Zoom* are self-explanatory, and you are encouraged to adjust them to see their effects. The *View Angle* controls how wide of a view we see. The wider the view, the more distorted the object looks.

Guided Tour 5.2

- Create a cylinder if you do not already have one on the screen. Switch to an isometric view of the cylinder. Display the cylinder with and without the hidden lines.

- Put the cylinder in a perspective view using *Views—Advanced—Perspective.*
- Try various setting for *Eye Distance* and *View Angle.*
- What happens when you *Repaint*?

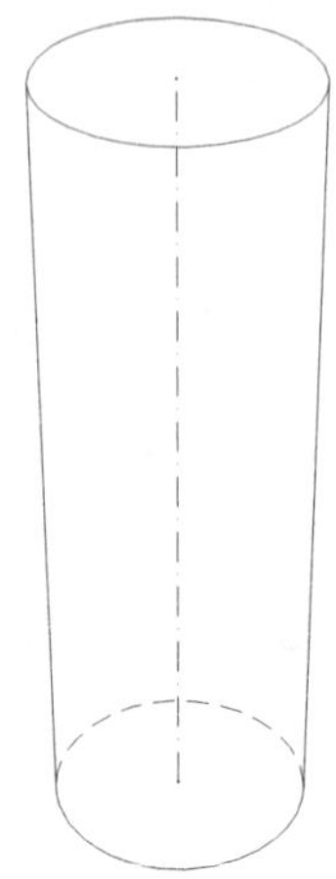

5.2.1 Object Edge Appearance in Pro/ENGINEER

On Pro/ENGINEER's shortcut tool bar, there are four options related to how you can display your object. The options include wireframe, hidden lines, no hidden lines, and shaded. Figure 5–6 shows the object appearance buttons from the tool bar.

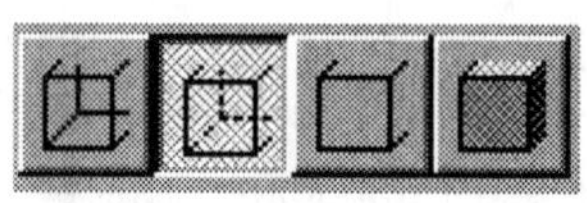

Wireframe I Hidden Line I No Hidden I Shaded

Figure 5-6 Object Appearance Short Cut Buttons

In Pro/ENGINEER, wireframe is the default object appearance setting. This allows a designer to easily see all the edges that make up the object. An object viewed in wireframe mode can be hard to interpret, as it is difficult to tell the front of the part from the back. For this reason, the hidden line appearance for the part is often preferred. Pro/ENGINEER is somewhat unique as its hidden lines are shown in gray. This limits the confusion that dashed lines can cause.

There are two object appearances which do not show any hidden edges, *No Hidden Lines* and *Shaded.* No hidden lines allows you to view only the frontmost surfaces without the clutter associated with the hidden surfaces. This appearance setting allows for quick regeneration, and dynamic viewing and is more "lifelike" when compared with wireframe or hidden line views. Shaded object appearance is discussed in the next section. Figure 5.7 shows the difference between the object appearance settings.

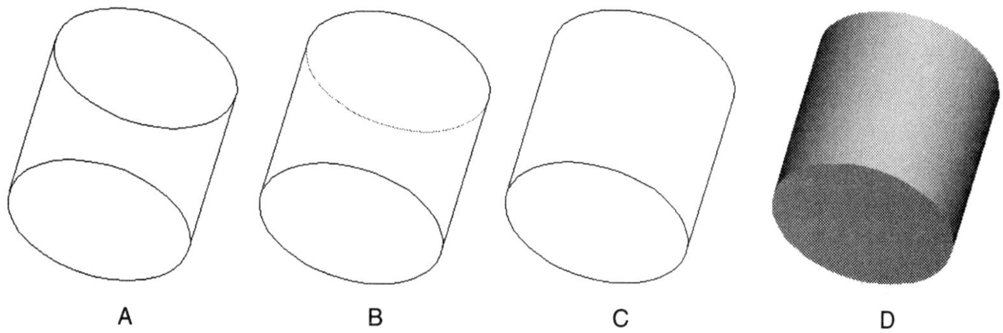

Figure 5-7 Examples of the Different Image Styles Available in Pro/ENGINEER

5.3 Object Surface Appearance

Surfaces are usually displayed as either shaded or not shaded. Some software packages call this *rendering*. All of the views shown in Figures 5–1 through 5–5 display the object with surfaces not shaded. When displaying an object with surfaces shaded, you can still use any of the view angle settings as described previously in this chapter. Depending on your software, you will probably have the option of showing your shaded image either with the edges visible or invisible. Displaying the shaded image with the edges invisible sometimes results in a hard copy that is unclear because the edges may not be well defined, based on the light source and the color of the object. Figure 5–8 shows shaded images of an object with and without edges visible.

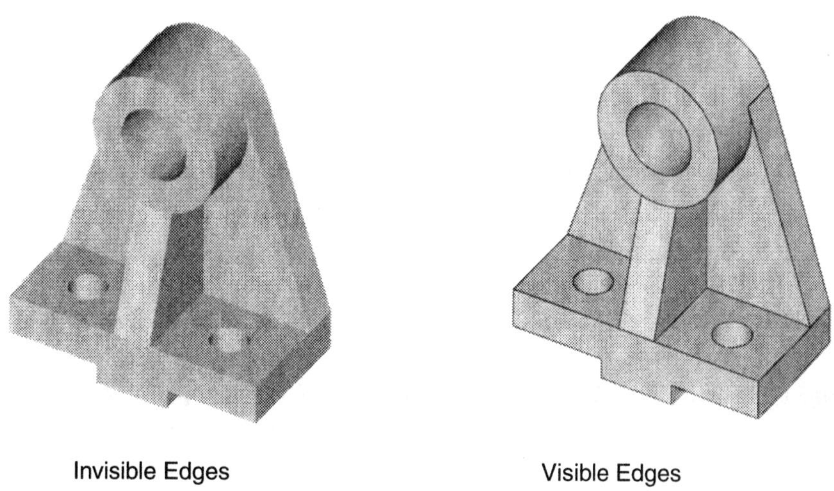

Figure 5-8 Shaded Surface Display

Another surface display setting is visible only when there are curved surfaces on the object. Curved surfaces exist on many objects, such as the cylindrical surface that exists in a circular hole. Sometimes, you may wish to show the curved surface with ruling lines or "meshes" on it. Although these lines do not appear on the actual surface of the object, their display can give a better definition to the surface. However, you may find that meshed surfaces are confusing Figure 5–9 shows an object with the curved surfaces displayed with and without ruled lines.

Curved Surfaces not meshed

Curved Surfaces meshed

Figure 5–9 Ruled Surface Display

One of the other surface display settings available in CAD software is the definition of the object color. Normally, you will be able to set the entire object to one color, or you may set individual colors for each surface on the object. You may not be able to plot your snapshots on a color printer, however, if you are creating shaded images to send to a grey-scale printer. You may find that lighter colors work best for you.

5.3.1 Surface Appearance in Pro/ENGINEER

When you select the shaded object appearance button from the shortcut tool bar (Figure 5-6), your part should look more realistic. With the shortcut tool bar, it is easy to toggle back and forth between the different object appearances. You may want to customize the appearance of your part by adding visible edges to the shaded view or by adding a meshed surface. Both of these options are found under the *View* pull-down menu.

Shaded View with Visible Edges Customizing the appearance of your model such that it appears shaded with visible lines requires you to use the *View* pulldown menu. The selection of *View—Model Display* will bring up a new dialog box as in Figure 5.10. From this box, you can select the from three categories, *General, Edge/Line,* or *Shade. General* repeats the shortcut buttons for object appearance. *Edge/Line* allows you to change the quality of how edges appear on the screen. The way tangent edges are viewed can also be modified in *Edge/Line*. To create a shaded image with visible edges, select the *Shade* category and then *With edges.* You can also adjust the quality of the shaded image; 10 is high quality, and 1 is low quality. Most of the other options found in this dialog box are highly specialized and beyond the scope of this text.

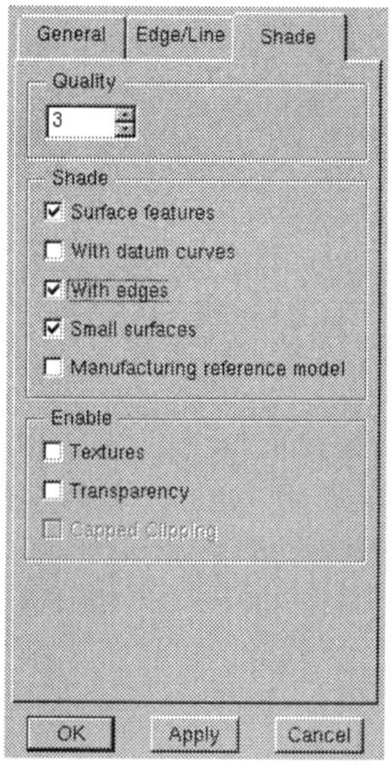

Figure 5-10 Model Display Dialog Box

Meshed Surfaces Applying a meshed or ruled surface to your object is easy to do in Pro/ENGINEER. To apply meshing to a surface, select the *View* pulldown menu, then *Advanced—Mesh surface.* A new dialog box should appear, and you simply click on the surface of interest to place the meshing. Once the surface has been "meshed" you can easily customize the mesh spacing using the entry boxes labeled *Mesh Spacing.*

Changing the Model Color When working with many different models that may eventually be pulled together on one drawing, it is advantageous to change the color of your model. This is done by selecting *View—Model Setup—Color&Appearance.* When the selection is made the Appearance dialog box appears (as shown in Figure 5–11). Pro/Engineering does not allow you to edit the default color setting, so you will have to add a new one. After you select *Add,* the Appearance Editor (Figure 5–11) pops up. We will keep this in Basic mode. To edit the color, simply click on the white box as indicated in Figure 5–11. This brings up the Color Editor, where colors are changed by sliding the controls back and forth. It is worthwhile to experiment with the different options found under *Color&Appearance*

In general, the display settings that you use when plotting your drawing will be different from those you use while working on your model. When you are working on your model, the amount of time it takes the computer to regenerate the screen to redisplay your object will vary depending on the complexity of your display settings. Wireframe display is the most economical in terms of time, that is, it will redisplay the fastest. Shaded images will usually take much longer to redisplay since the computer must perform numerous computations taking into account the light source relative to the view angle before it can redisplay the computer screen. Display of the object with hidden lines invisible will fall somewhere between these two in terms of the amount of time necessary to regenerate the screen. If you are working on a fairly simple model, the differences in display

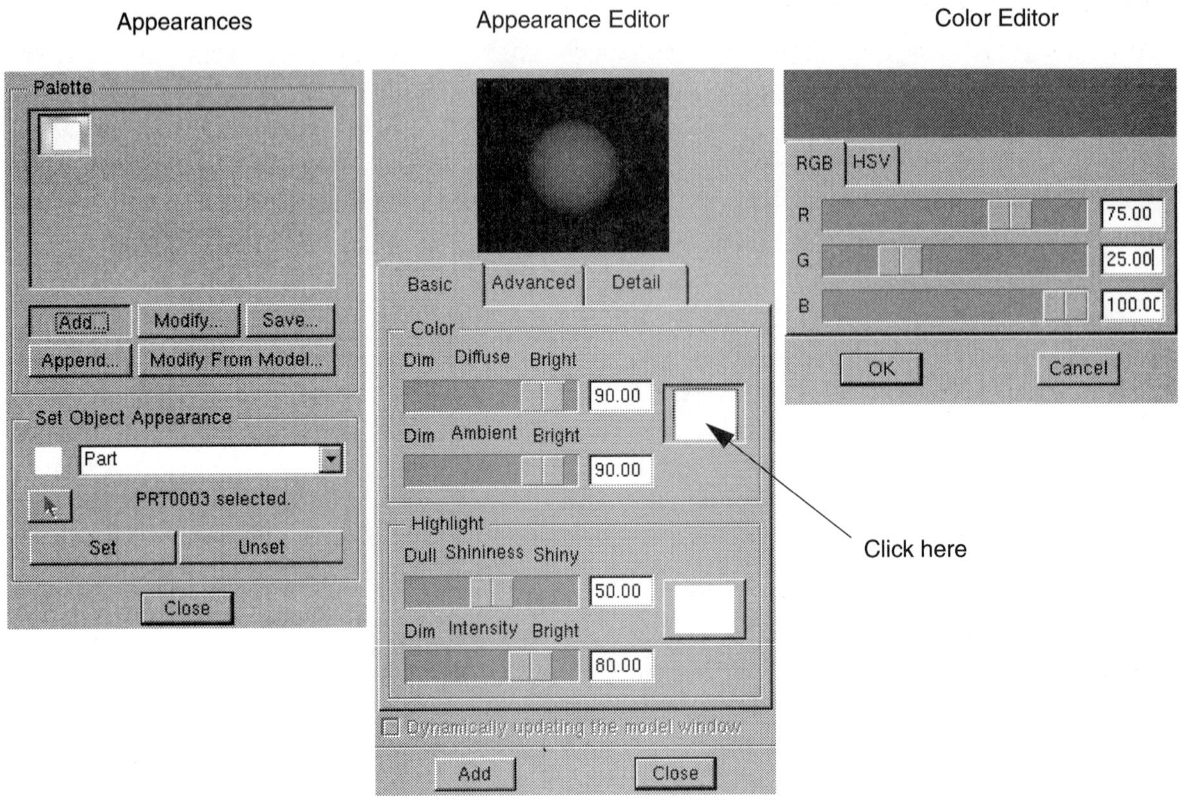

Figure 5–11 Appearance Dialog Boxes

time may be small. However, if you are working on a complex object with many surfaces and edges, you may find that your ability to work productively is hampered by anything other than a wireframe display.

Guided Tour 5.3

- Retrieve or create the cylinder used in Guided Tour 5.2.
- Display your model with shaded surfaces.
- Modify the color of the model to the setting found on Figure 5-10
- Select *View—Model Setup—Light* and create a point source light source. Experiment with different setting for longitude, latitude, distance.

5.4 Display of Nongeometric Object Entities

Along with the edges and surfaces that make up your object geometry, there are other entities associated with the geometry that will make your life easier as you create and edit objects. These entities include datum planes, axis lines, datum points, and the coordinate system. Sometimes you will want to display or hide these entities as you create drawings or plots of the model. Figure 5–12 shows an object viewed with hidden lines and various

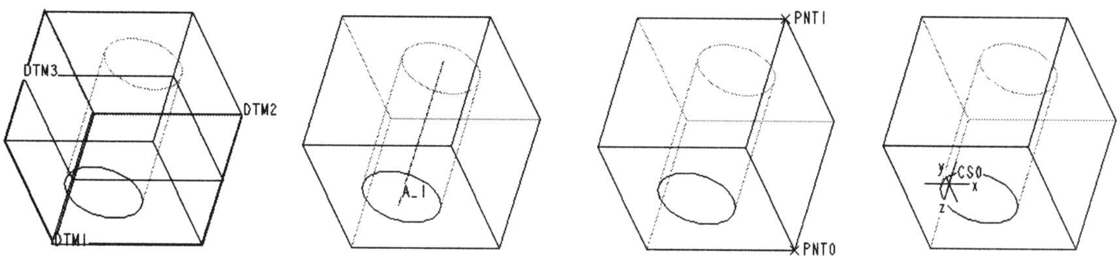

Datum Planes | Axis | Datum Points | Coordinate System

Figure 5–12 Nongeometric Entities Display

nongeometric entities displayed. Note that these nongeometric entities are a permanent part of the model definition, and turning off their display does not delete them from the geometry, but only hides them temporarily from view.

5.4.1 Display of Nongeometric Entities in Pro/ENGINEER

In Pro/ENGINEER, it is important to note that you must create something in order to display it. This holds true when displaying nongeometric entities as well. Earlier chapters discussed the procedures for creating datum planes, coordinate systems, datum points, etc. The creation of these entities is usually done prior to creating any geometric features of the part. Once created, the nongeometric entities are easily viewed or hidden with the use of the shortcut tool bar. Figure 5–13 shows the various nongeometric entities which can be turned on or off.

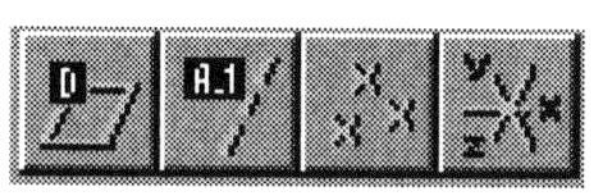

Datum Planes I Axis I Points I Coordinate Systems

Figure 5–13 General Nongeometric Entities Shortcut Buttons

In Sketch mode, four additional nongeometric entities shortcut buttons appear. These buttons toggle the dimensional values, constraints, background grid, and section vertices. Figure 5–14 shows the push buttons for these operations.

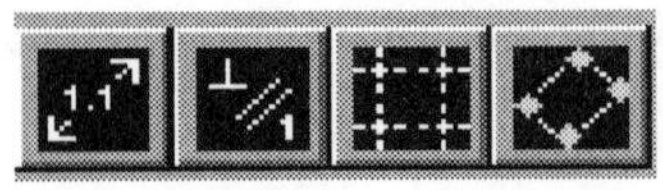

Dimensions | Constraints | Grid | Section Vertices

Figure 5–14 Sketch Mode Nongeometric Entities Shortcut Buttons

---

Guided Tour 5.4

- Start a new part, create a set of datum planes and a coordinate system.
- With the Intent Manager turned on, create a block with a circular hole through the entire center of the part (see the following figure) .

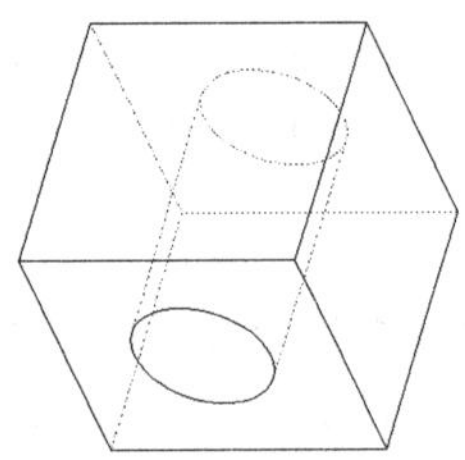

- Toggle each of the nongeometric entities on and off.
- What happens when the datum plane is on, and the model is displayed in shaded view?

5.5 Dynamic Viewing

Most modern CAD software packages allow you to set your desired observer location display dynamically. This means that you can zoom or rotate the object in real time on the computer screen instead of using the preset viewing options. Dynamic viewing is a powerful tool that you can use during the design and modeling process. Depending on your available hardware, you may be able to use dynamic viewing with shaded images or with invisible hidden line settings, however, many times you will only be able to use dynamic viewing options with wireframe display.

5.5.1 Dynamic Viewing in Pro/ENGINEER

Dynamic viewing in Pro/ENGINEER is accomplished by holding down the control key and using one of the mouse buttons. Depending on your graphics card, you may be able to do dynamic viewing with a shaded image on the screen, but chances are you will only be able to do dynamic viewing with a wireframe or line display of the object. The function keys of particular interest in dynamic viewing are described in the following:

Right Mouse Button and Control Key Hold down the keys and move the mouse to translate the view of the object in like manner. This is known as *panning* the workspace. For example, moving the mouse to the right results in the object appearing to move to the right on the screen (remember that the object is not really moving—your viewpoint of the object is changing).

Left Mouse Button and Control Key Hold down the keys to zoom in or out. Sliding the mouse *down* the mousepad (towards yourself) will *zoom in* on the object; sliding the mouse *up* (away from yourself) will *zoom out* from the object.

Middle Mouse Button and Control Key Hold down the keys to dynamically rotate your view of the object. The object will appear to rotate in space, however, remember that the object is fixed in space and that you are moving around it.

To return to the default view, you can select *View—Default* or use one of the shortcut tool bar buttons.

Guided Tour 5.5

- Create a simple part of your choice. Experiment with the dynamic viewing options by holding down control key and the left, middle, or right mouse button and moving the mouse on the mousepad.
- Try the dynamic viewing operations with the line option, the hidden lines removed option, and the shaded option as your starting point.

5.6 Troubleshooting/FAQs

Problem	Potential Solution
You turned the hidden lines off, but they appear on your plot.	Select the *Repaint* shortcut button.
Shaded image is too dark on the plot.	Try changing the color of the object to either pink or yellow (something with a lighter shade).
Shaded image has no definition on the plot—you can't tell where one surface begins and the other one ends.	Make sure you select a color for the outline that contrasts with the color of the object itself. You may also need to modify the lighting options.
Plot/graphics region contains unwanted nongeometric entities.	Remember to turn off nongeometric entities prior to printing.
You sent a plot of a shaded image to the printer, but nothing seems to be printing.	Shaded image plot files are typically several orders of magnitude larger than plot files from line displays of the object. Either you should be patient—the plot will be coming out soon, or you have exceeded the memory capacity of the printer in the lab. If you seem to have exceeded the memory capacity, talk to a lab instructor for help.

5.7 Chapter Summary

In this chapter, you learned about basic 3-dimensional display techniques. You learned about how your view of an object depends on the observer location as well as the way you choose to display edges and surfaces of the object. You learned how to control the view angle, how to control the edge and surface display, and how to set the desired visibility of nongeometric entities. Finally, you learned how to dynamically view your object in real time by panning, zooming, or rotating.

Exercises

On a clean sheet of paper, develop a **Plan of Action** for the object assigned by your instructor describing the steps you would follow to create the object in Pro/ENGINEER. Use sketches as appropriate in your Plan of Action. Then create the object in Pro/ENGINEER. Create your parts to the exact sizes of the features on the object. Obtain a view of the object specified by your instructor and turn in hard copy as appropriate.

1. Create the following part. Add a surface mesh to all curved surfaces.

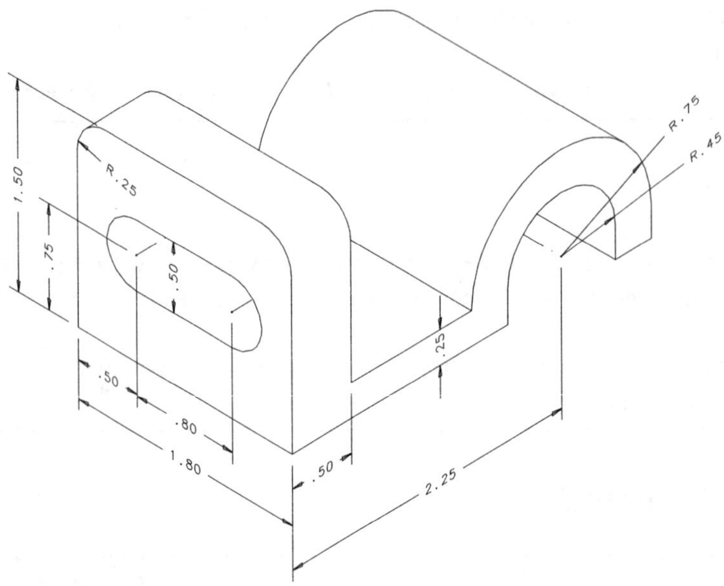

2. Create the following part. Once completed generate a perspective view of your part. What setting were used to obtain your perspective distortion.

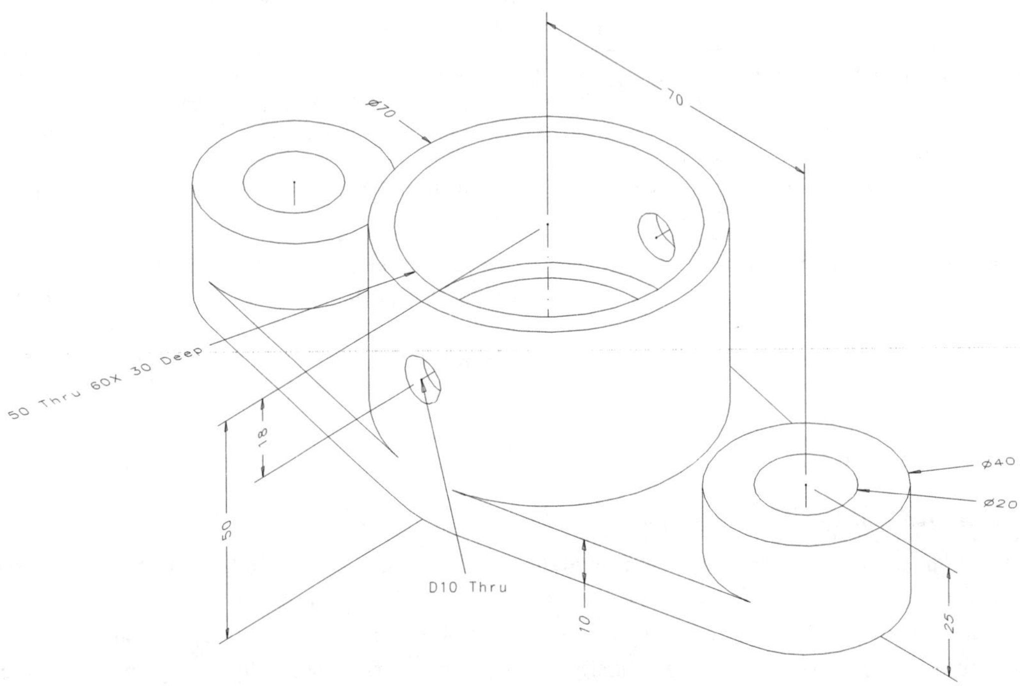

3. Create the following part. Apply a surface mush the part's underside.

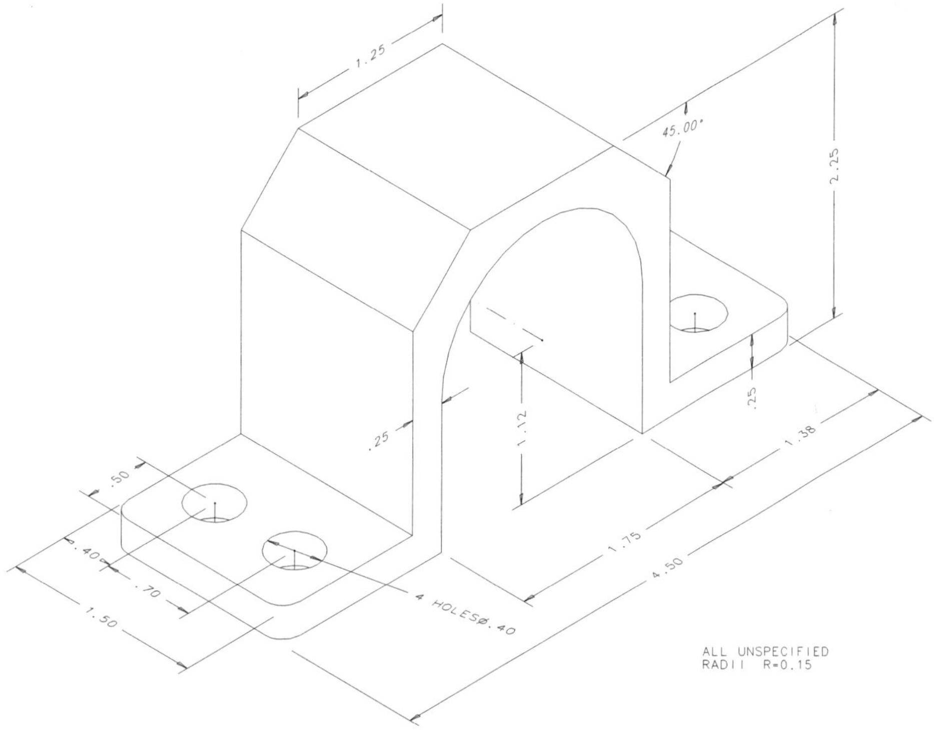

4. Create the following part. Apply a mesh to all curved parts. Generate a perspective view of this part.

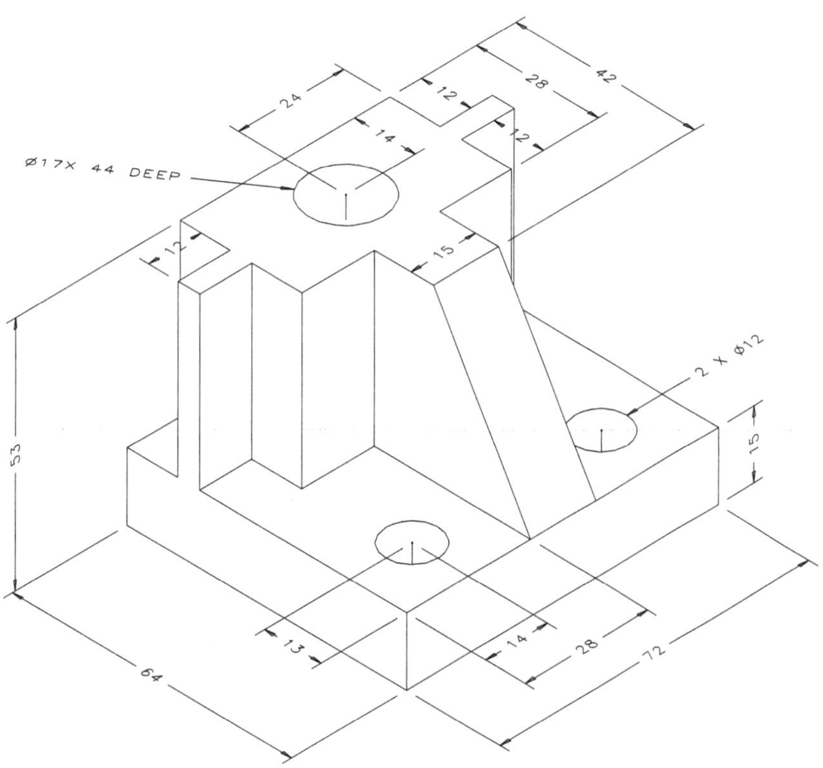

5. Create the following part. Once completed generate a perspective view of your part. What setting were used to obtain a "normal" looking perspective?

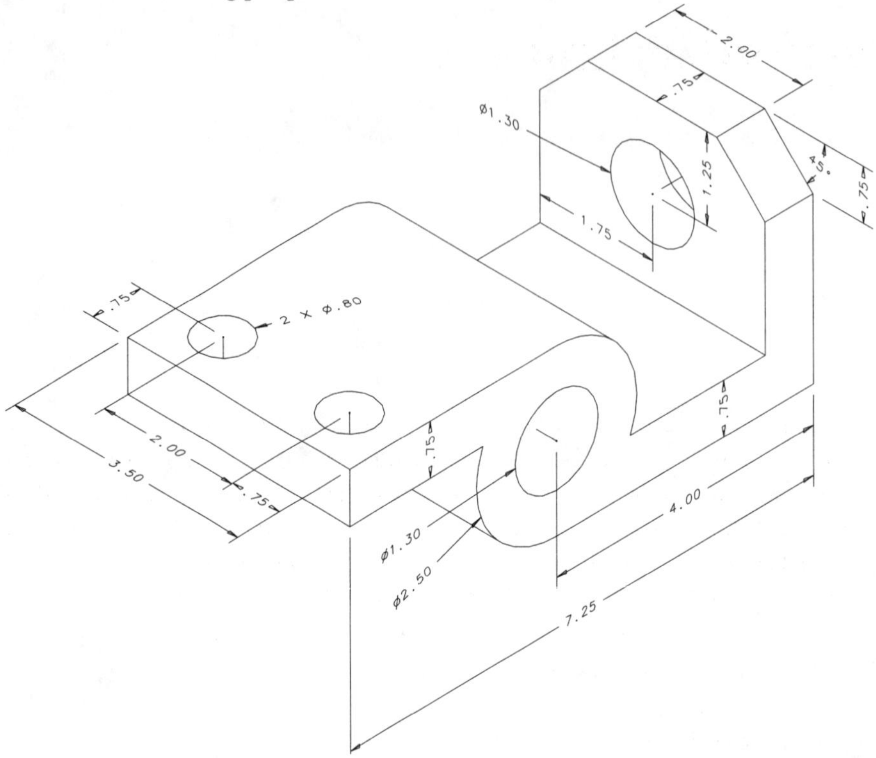

6. Create the following part. Once completed generate a perspective view of your part. What setting were used to obtain a "normal" looking perspective?

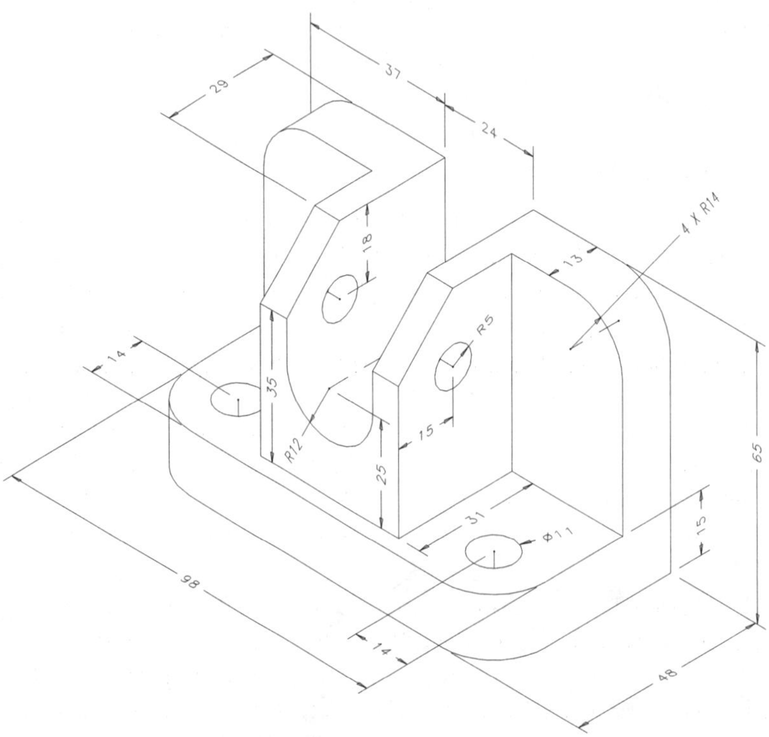

CHAPTER 6

Construction Techniques

As stated in Chapter 1 of this text, computer-generated objects are typically made up of features. Most modern CAD software employs a methodology know as "feature-based modeling." The features combine with objects or other features to create the desired final object. The features available in the software are usually designed to relate to how engineers think in their design and manufacturing work. Most software packages have predefined features that are used to build objects, one feature at a time. As you create the specific features needed to define your required object, you supply the dimensions and the location of the features relative to the edges, points, and surfaces on the object. In this chapter, you will learn how to use construction techniques focused around Pro/ENGINEER's use of reference datum planes.

6.1 Feature-Based Modeling

As stated previously in this text, most objects can be thought of as being made up of several features. The features that make up objects are designed to relate to "how engineers think" about part geometry. In this way, they are tied closely to the design and manufacturing thought process. With most modern feature-based modelers, you have the ability to create your own features or to use the predefined features in the software. You supply the actual dimensions for the parameters of the feature geometry to create the feature to your specific size requirements.

6.1.1 Standardized Part Features

In addition to primitive shapes, many solid modeling software packages contain other **standard features** for object creation. These features are readily available with predefined shapes for sizing and can be used in either a cutting or a joining operation. An example of this type of feature is the counterbored hole feature shown in Figure 6–1a. This feature is used to create a counterbored hole in an object. A block that has this feature removed from it is shown in Figure 6–1b. The counterbored hole feature was used in a *cutting* operation on the object. The parameters used to create the counterbore feature are shown as dimensions in Figure 6–1a and consist of the diameter and depth of the drilled hole and the diameter and depth of the counterbore. Notice that the entire depth of the drilled hole is given ("3" in the figure), not just the height of the smaller cylinder by itself. This is in keeping with the manufacturing process used to create a counterbored hole where the smaller diameter hole is first drilled to a specified depth, then the larger diameter counterbore is machined to a specified depth.

The counterbored hole feature could have been created as a separate object for cutting, but you would need to create a separate cutting object composed of two joined cylinders each time you wanted to create a counterbored hole in your object. You would also need to create a counterbore cutting object for any future objects you were creating that contained counterbored holes. By including a counterbore as a built-in cutting feature within the solid modeling software, the amount of time spent in generating the model is greatly reduced, and the design is much more flexible. Other cutting features that are routinely included in 3-D modeling software are

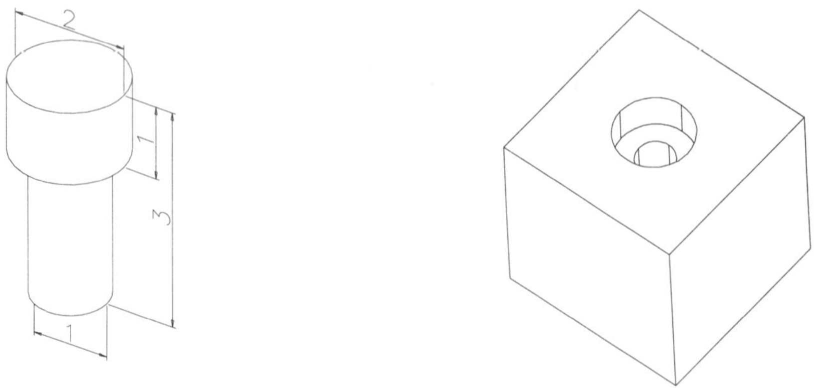

Figure 6–1 Counterbored Feature

countersunk holes and drilled holes. Figure 6–2 shows the top and front views of a block that includes counterbored, countersunk, and drilled hole features.

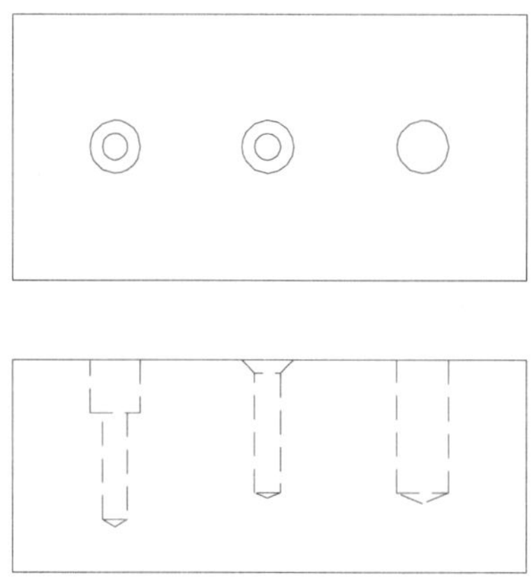

Figure 6–2 Counterbored, Countersunk, and Drilled Hole Features

Ribs are thin geometric entities that are often added to an object to strengthen it or to give it structural stability. If available in the software, ribs are another type of feature that can be easily added to a model. They are included in an object as a *joining* feature, unlike the cutting features described for counterbored, countersunk, or drilled holes. Figure 6–3 shows an object with a rib feature added to it as well as the geometric parameters that were used to define the feature. In this case, the defining parameters were the rib thickness, the angular orientation of the rib, and the length of each leg of the rib feature.

Chamfers are beveled edges that exist on many engineered parts where the sharp edges are removed for some reason. These features are available in many software packages. You have probably seen a chamfered edge at the end of a bolt if you recall that the circular end of the bolt was beveled to eliminate a sharp edge. Another type of feature that is readily available in most solid modeling software is a fillet (or a round). A rounded interior corner is called a "fillet" and a rounded exterior corner is called a "round." In general, sharp corners are avoided in designing parts to be cast or forged because they are difficult to produce and because

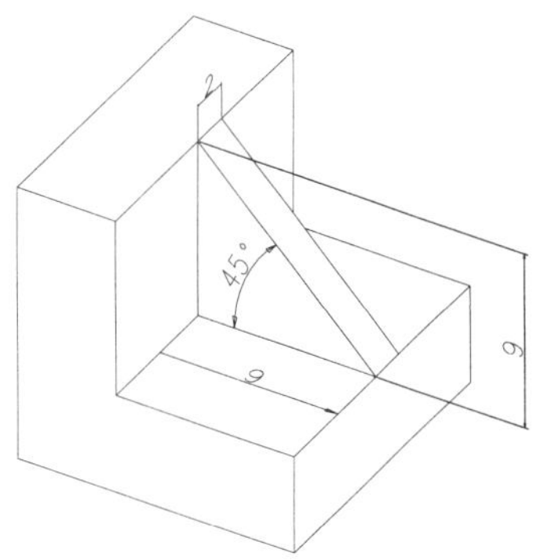

Figure 6–3 Object with Rib Feature Joined

they are a source of stress concentration, particularly for interior corners. For this reason, sharp corners in objects are subject to failure due to the relatively high stresses in their vicinity.

Depending on the software you are using, you may have to be careful when adding fillets, rounds, or chamfers to object geometry. It is often desirable to add the fillet or chamfer features to an object after the basic object geometry is established because the way that fillets, rounds, and chamfers interact at a corner may be adversely affected by subsequent operations on the part geometry. Figure 6–4 shows the front and isometric views of a simple part that contains chamfered and filleted edges.

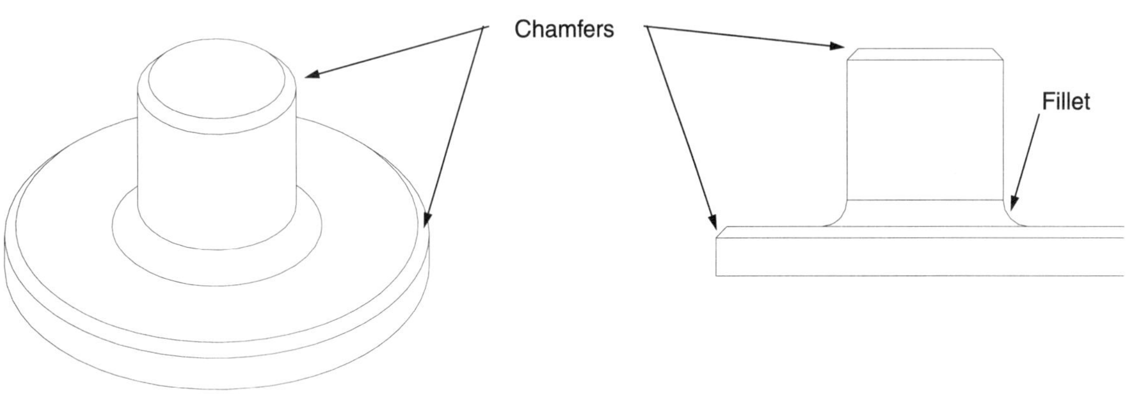

Figure 6–4 Chamfer and Fillet Features

6.1.2 Locating and Constraining Features

In general, features will have to be located on the object with respect to datum planes, object edges, or other well-defined points on the object. Feature locations are similar to constraints as described in previous chapters. For example, if you are specifying the location of a drilled hole, you need to specify the location of the circle center relative to the object edges. (This is in keeping with the manufacturing process of locating a drilled hole on an actual part.) With feature-based modeling, you could choose to locate the hole center as a certain with unchanging offset from an edge. Alternatively, you could choose to locate the hole center as a percentage (say

50%) of the distance along an edge. Either way, if you change the dimensions of the part, the location of the hole will be updated to reflect the change in part geometry.

Figurc 6–5a shows thc top vicw of a block which contains two drilled holes. The holes are located on the block such that the center of the first hole is a fixed distance of 1.0" from edge A and a variable distance of 50% of the length of edge B. Figure 6–5b shows the same object after the lengths of edge A and edge B have been modified. Note that the location of the holes relative to the edges has remained the same even after the object was modified.

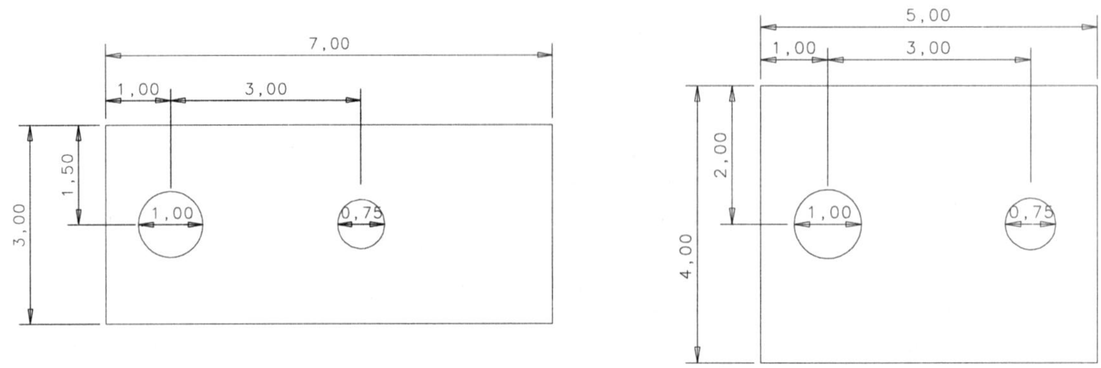

a. b.

Figure 6–5 Part with Constrained Features

Feature locations and sizes can also be constrained by the relative sizes of other features on the part. For example, the block in Figure 6–5a locates the center to center distance between the two holes (in the direction parallel to edge B) equal to three times the diameter of hole C, and the diameter of hole D is constrained to be equal to 75% of the diameter of hole C. Thus, if the diameter of hole C is changed, the distance between the two holes will change, as will the diameter of hole D. Figure 6–6 shows the block from Figure 6–5a after the diameter of hole C has been modified to be 1.5 inches instead of 1.0 inch, as originally designed.

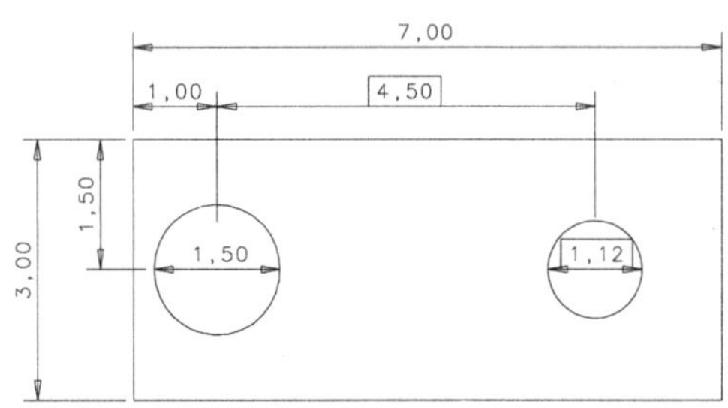

Figure 6–6 Modified Part Geometry

6.1.3 Adding Standard Features in Pro/ENGINEER

Several standard features are also available in Pro/ENGINEER, and we will continue to focus on holes, rounds, chamfers, cuts, protrusions, and ribs. There are other features, but they are beyond the scope of this text.To access the available features in Pro/ENGINEER, select *Feature—Create,* then choose the feature of interest. Figure 6–7 shows the available standard features which can be created using Pro/ENGINEER.

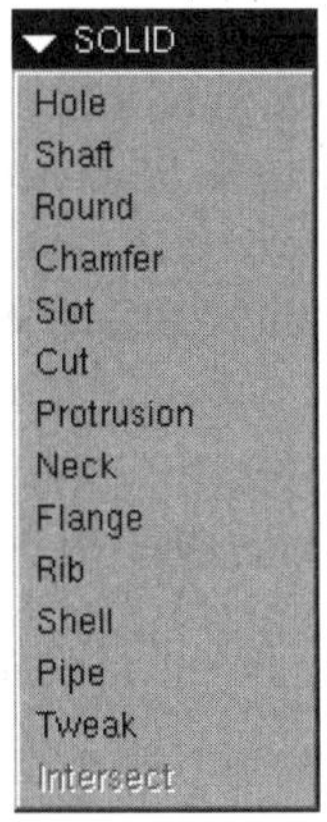

Figure 6–7 List of Available Features

Each feature is sketched on a datum plane or on a surface of the part. Once the surface of interest has been selected, you must define the direction in which the feature is to be added, then sketch and constrain the section. Table 6–1 helps to summarize the required parameters for each feature. Explore how each of these features is added to your part.

Table 6-1 Summary of Required Parameters

Feature	Shape	Required Parameters	Helpful Hints
Hole	circular	location, size, material side, depth	
Round	round/fillet	edge or edge chain, radius	avoid large and overlapping round
Chamfer	linear/flat	edge or edge chain, angle, and size	avoid overlapping
Cut	user defined	shape, location, material side, depth	
Protrusion	user defined	shape, location, material side, depth	
Rib	user defined	shape, location, material side thickness	create an offset datum plane as your sketching plane

6.1.4 Rounds and Chamfer Features in Pro/ENGINEER

Fillets and chamfers are added to your three-dimensional geometry as separate features. Sometimes, fillets, or chamfers will be included in your 3-D geometry because you created a 2-D shape that had these features in it and either extruded or revolved the shape. However, if you create a 3-D part with simple geometry, you may need to add fillets or chamfers to the object geometry itself. To add a fillet to a 3-D part, select *Feature—Create—Solid—Round.* For the 3-D geometry, fillets are know as rounds.

After you have followed the command sequence for rounds, you will be prompted to choose the edges, vertices, or surfaces to fillet and round. At this point, you can cursor select any edges or corners where you wish to locate a fillet or round. If you select a single edge, only that edge will include a fillet. You will then be asked to supply the desired fillet radius (you can also choose to include a fillet of variable radius). Once you input the fillet radius, press *Return* one time, and you will continue to be prompted to select the edges or vertices to fillet. When you are finished specifying all of the fillets for your part, you need to press the *Return* key one extra time. By doing this, the fillets will be included in your part all at one time. This is desirable because if you were to create the fillets one at a time, the corner intersections would not be "smooth." Therefore, you should specify all of the desired edges for filleting and let Pro/ENGINEER figure out how to transition the fillets at the intersections. Figure 6–8a shows a block with three filleted edges that meet in a corner. In this case, the edges were filleted one at a time. Notice that these three fillets meet in a sharp corner on the part—not the desired result. Figure 6–8b shows a similar block, only this time the three edges were filleted simultaneously, and the software determined how to best fit a smooth surface between the rounded edges.

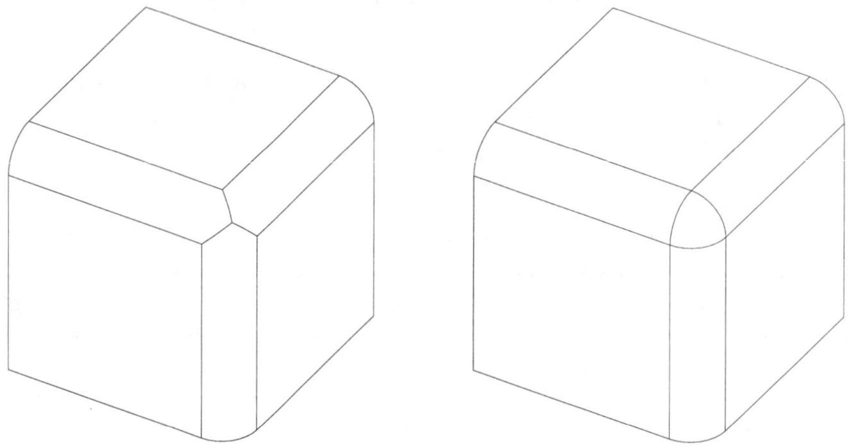

a. Edges Rounded "One by One" b. Edges Rounded by "Tangent Chain

Figure 6–8 3-D Rounds on an Object

To create a chamfered edge, select *Feature—Create—Solid—Chamfer.* You will be prompted to select the edges or vertices to chamfer and then to supply the scheme offset to use. Most of the time 45° x d will be acceptable. Once again, you should chamfer all of the edges that you intend to before pressing *Return* the final time. Pro/ENGINEER will then determine the transitions between any intersecting chamfers.

It is best to save all of your filleting and chamfering operations until the end when you are creating your 3-D parts because they will look strange at their intersections if you do them one at a time. However, if you mistakenly add fillets or chamfers before you are finished creating your solid model, you can delete these features from your object definition and redo the filleting and/or chamfering at a later time. Deleting features from your part definition is covered in Chapter 7 of this text.

Guided Tour 6.1

- Switch the units to inches.
- Create a block with height = 2, length = 3, and depth = 2.25. Cut a section with height = 1, length = 2.25, and depth = 2.25 to form an "L" shape.

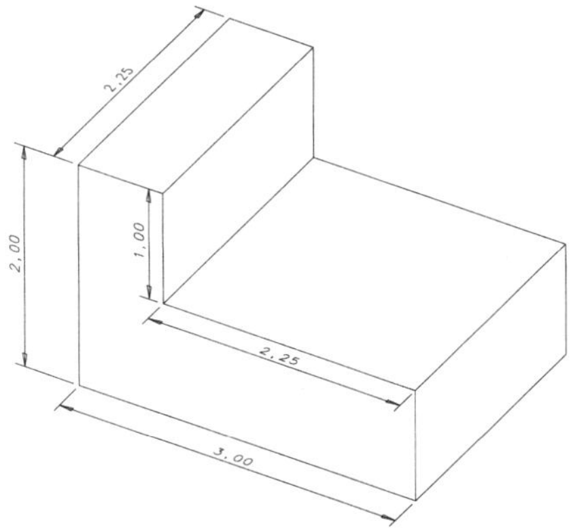

• Add a cut feature with height = 0.5, length = 1, and depth = 2.25 to obtain the staircase shape shown in the first two figures of this Guided Tour.

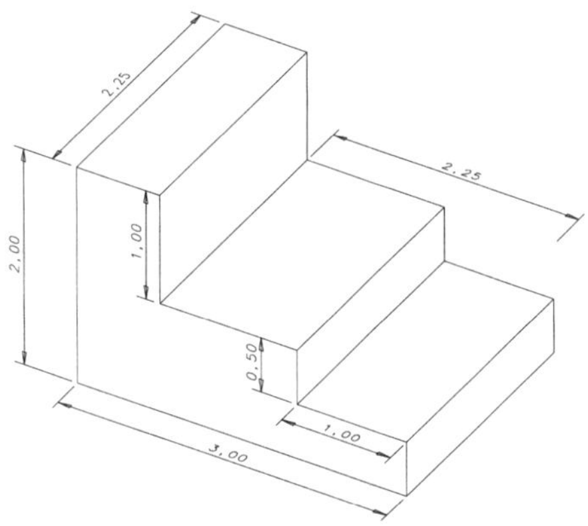

• Create a small cut with dimensions height = 0.4, length = 0.8, and depth =1.25. Cut this shape from the top of the staircase.

• Add a Hole feature with radius = 0.25, depth = 1. Create the hole using the linear placement so that it is 0.75 and 0.5 from the indicated edges. When you are finished, your object should look similar to the one shown in the following figure:

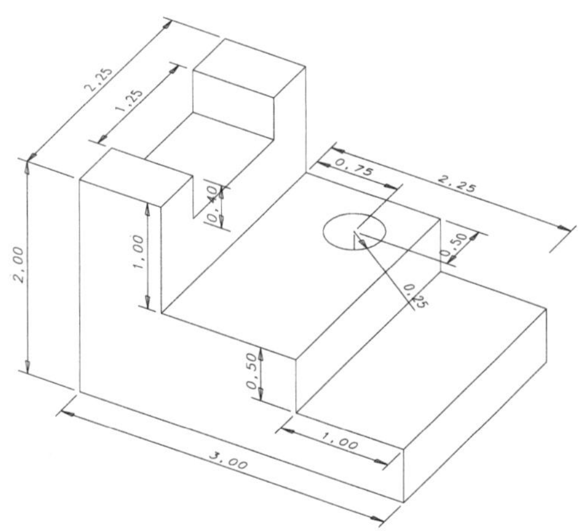

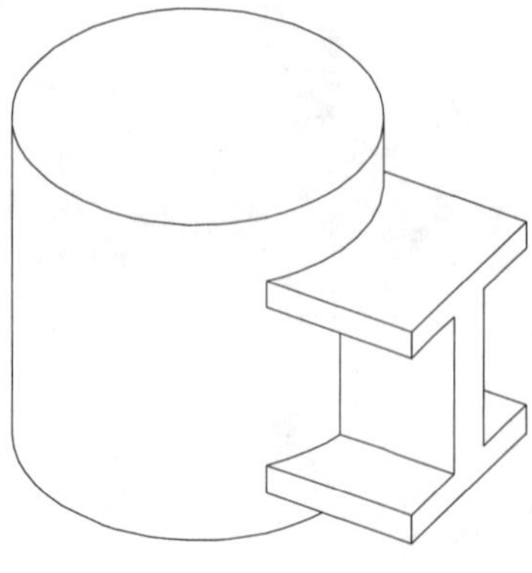

Figure 6–9 Object from Cylindrical Post and Cross Bar

6.1.5 Reference Geometry

Sometimes when you are combining two objects, you will not have a planar surface for sketching or for orienting features and adding relationships. Suppose you want to create the object shown in Figure 6–9. You could easily create the two separate features, locate them in space, and join them. This would result in the required object, however, you could not easily modify the object at a later time and end up with the desired part geometry. Because the cylindrical post has a curved surface, you can not create the desired dimensional relationship between the end of the crossbar and the center of the shaft. This can be remedied by adding a *reference plane* to the geometry of the cylindrical post feature so that you have a flat surface for orienting the feature of the crossbar. Figure 6–10 shows the cylindrical post with the reference plane added to it. Notice that the reference plane can be oriented with respect to the size features of the cylinder or relative to the coordinate axes for the system. The crossbar can be accurately sketched on this reference plane as shown in Figure 6–11 and the section extruded to form the part shown in Figure 6–9. You now have a relationship between the post and the crossbar through the use of a reference plane. If you modify the dimensions of either the cylindrical post or the crossbar, these relationships will be maintained.

In addition to reference planes, you can also add reference points and reference lines to the part geometry. This reference geometry becomes a permanent part of the object definition. Figure 6–12 shows a block with two countersunk holes drilled from its front surface. Note that for this object, a reference line was created on the

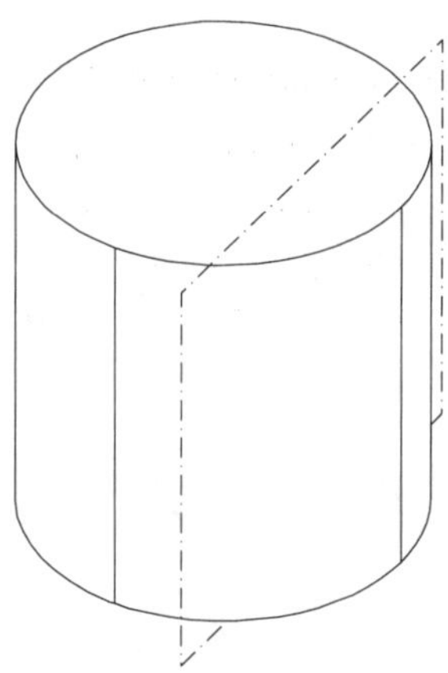

Figure 6–10 Cylindrical Post with Reference Datum Plane

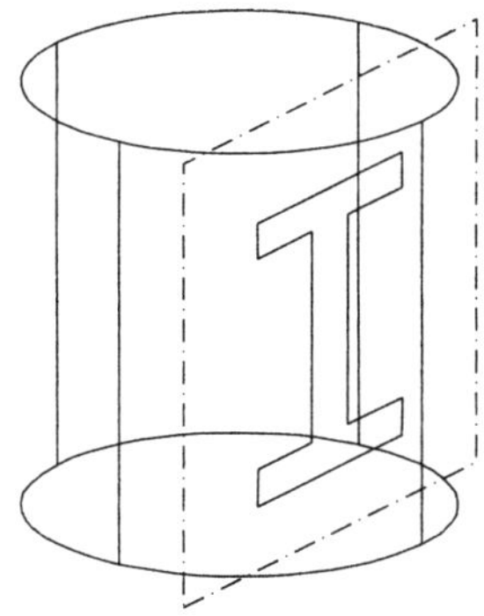

Figure 6–11 Sketch on Reference Datum Plane

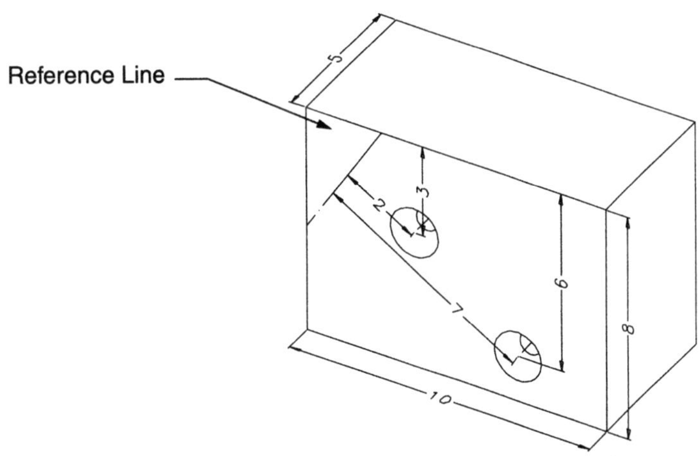

Figure 6–12 Object with Reference Line Included

front surface of the block, and each of the countersunk holes was located relative to this reference line as well as to one of the edges of the block.

6.1.6 Datum Plane Operations in Pro/ENGINEER

As you may have realized by now, the reference datum plane plays an important role in the development of objects using Pro/ENGINEER. The datum planes serve as reference planes on which to sketch or from which to measure. In most cases, you will start a new part by first creating a set of "default" datum planes. Once these are created, further datum planes can be created as the need arises. These planes become a permanent feature of your part.

***Offset* Datum Command** Not counting the default datum planes you have already worked with, the offset datum plane is the easiest to create. After selecting *Create—Datum—Plane—Offset*, you will be prompted to select a surface from which to offset. You then select any planar surface that is visible on the screen. Once the plane is highlighted, select *Enter Value*. You will be prompted to input the offset distance. If you input a positive number, the offset will be in the direction of the arrows for the datum plane. If you input a negative number the offset will be on the opposite side of the plane.

***Through* Datum Command** An extremely useful way to create inclined or oblique datum planes is the use of the *Through* command. This command is used to create a datum plane through *Axis/Edge/Curve*, *Point/Vertex*, *Plane*, or *Cylinder*. Recall that three non-colinear points are required to define a plane. For this reason we need to use the *Through* command in the selection process multiple times. If three points are to be used to define the plane, we must step through the selection three times. If however, a line (edge, cylinder, etc.) and a point are to be used to define the plane, we need only step through the selection twice. Figure 6-13 shows the points selected to create an oblique datum plane.

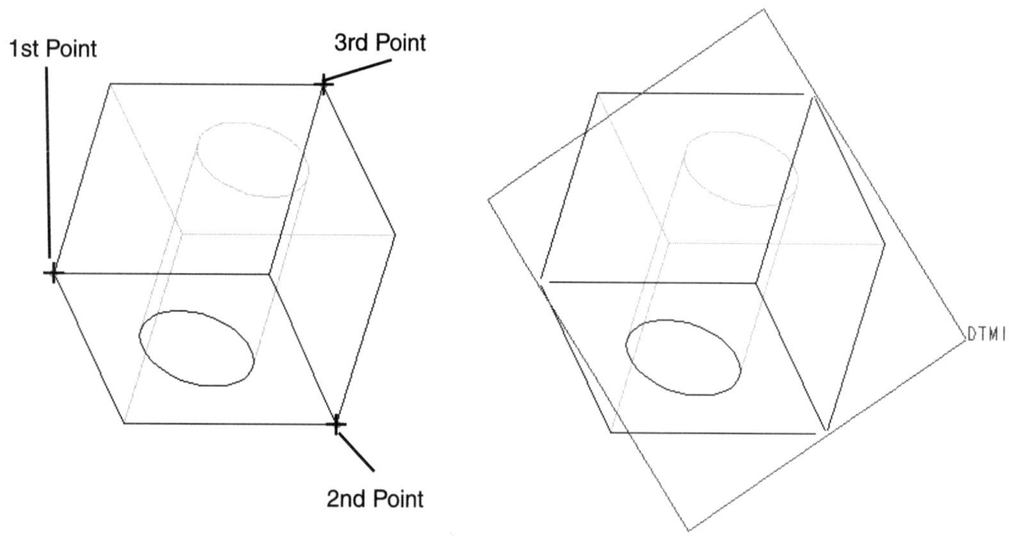

Figure 6-13 The Creation of an Oblique Datum Plane Using the *Through* Command

***Normal* Datum Command** The third method used to generate datum plane is the *Normal* command. This creates a datum plane set at 90 degrees from another surface. This is commonly used following the creation of an oblique datum plane to create a complementary oblique plane. To define a normal datum plane you must first define a reference plane from which the new plane will be normal, and a point or edge the new plane will pass through (this uses the *Through* command). Once you know how to create addition datum planes, you are ready to make complex objects.

Guided Tour 6.2

- Start up a new part and make sure millimeters are specified as the default units. (Do NOT create a default set of datum planes.)

Set up—Units—Principle System

- Create a part similar to the one shown in Figure 6.13, 100 mm on a side with a hole of diameter 50 mm.

- Create an oblique datum plane as in Figure 6.3 using the *Through* command (Note: you will have to "unselect" *Axis/Edge/Curve, Plane,* and *Cylinder*).

- Create a plane normal to the oblique datum plane through points "A" and "B."

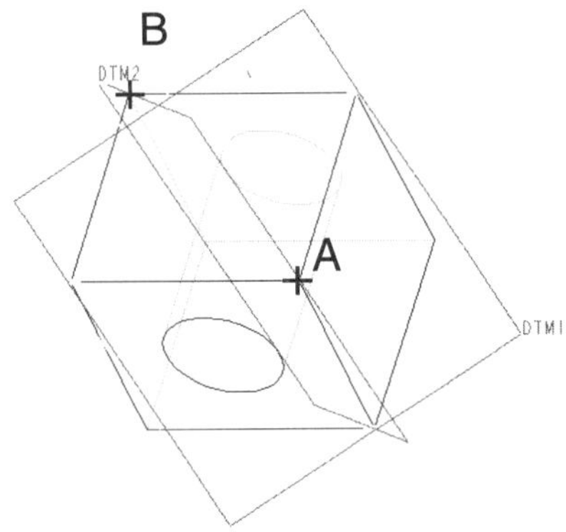

6.1.7 Adding Other Reference Geometry in Pro/ENGINEER

You are able to include coordinate systems, planes, curves (lines), and points as reference geometry for the parts that you create in Pro/ENGINEER. As stated previously, the purpose of reference geometry is to enable you to include relationships between it and any features you may wish to add to your part. Therefore, if you attach a reference point to your part geometry, you can locate features relative to this point. Similarly, you can establish a reference coordinate system for a part that is different from the overall or global coordinate system. This may make it easier for you to work with your model, depending on the data you are working with at the time. The options used to add reference geometry to a part are all found under *Feature—Create—Datum*. Figure 6-14 shows the various options available for creating reference geometry.

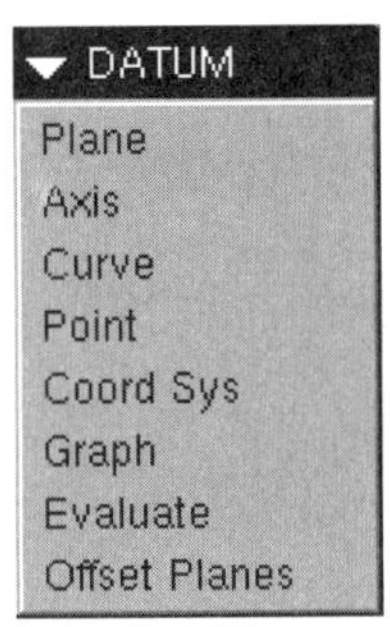

Figure 6-14 Available Reference Geometry

Reference geometry is very helpful when using the Intent Manager. Recall that the Intent Manager requires that you first select references from which to constrain added features. The reference geometry can be points, curves, planes, etc. We have already discussed how to add planes, so let's look at how we can add other reference geometries.

To add reference datum **points** to a part, you first select *Feature—Create—Datum—Point*. Here you will find a list of various options for placing or creating your part. For this text, we will focus on three options. *On Surface* allows you to select a surface on which you will insert your reference point. You are asked to select to edges or vertices from which to measure or constraint the location of your point. *On Vertex* allows you to put a point at a vertex or the intersections of planes. *On Curve* allows you to place a reference point on a selected curve. You can select *offset*, how far you want the point from one end, or *ratio*, a proportionate distance from one end (e.g., 0.50 represents half the distance of the curve).

Reference datum **curves** are easily added. The process of adding a reference curve is as simple as creating a basic sketch on a surface. The process of creating a reference line or curve starts using the *Feature—Create—Datum—Curve*. Some options which are more advanced than this text will appear, but we will stick to the *Sketch* option. Once *Sketch* is selected, select *Done* and then the sketching plane. The reset of the procedures is identical to adding any other geometric feature to your part.

Guided Tour 6.3

- Set the units to millimeters and create a block that is 100 mm x 50 mm x 25 mm.

- On the 100 mm x 50 mm surface, add a reference line as shown in the following figure. Select *Create—Datum—Curve—Sketch—Done.*

- Once in Sketch mode, select all four sides as references for the Intent Manager. Sketch the diagonal, as follows:

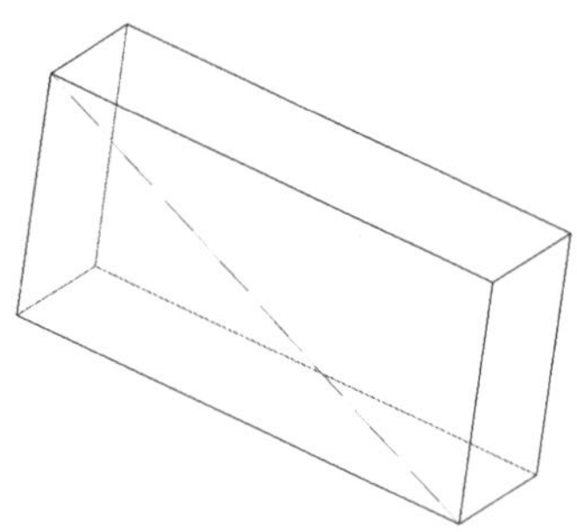

- Create three 10 mm diameter holes along the reference line (make sure to select the line as a reference prior to sketching).

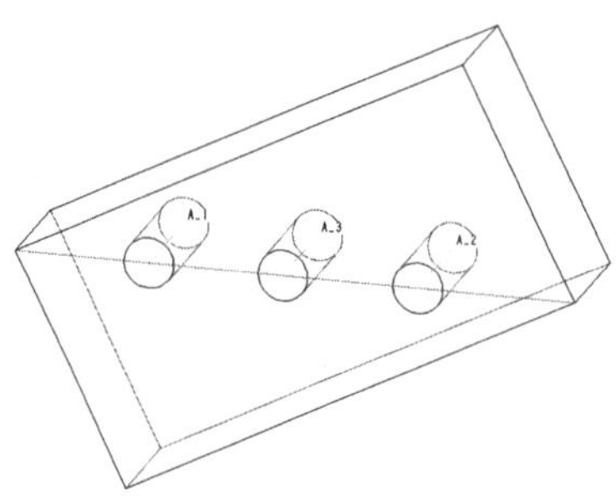

6.2 Construct Strategy

As stated previously in this text, there is usually more than one way to create a solid object using a computer. For example, the L-shaped object shown in Figure 6–15 could be created by drawing the basic L-shape as a two-dimensional figure and extruding this to form a 3-D object, or it could be created by starting with a large block and cutting a small block from it. In this section, we will focus on the strategies that have been described in this chapter of the text. You are free to experiment with other methods in the creation of these objects if you desire.

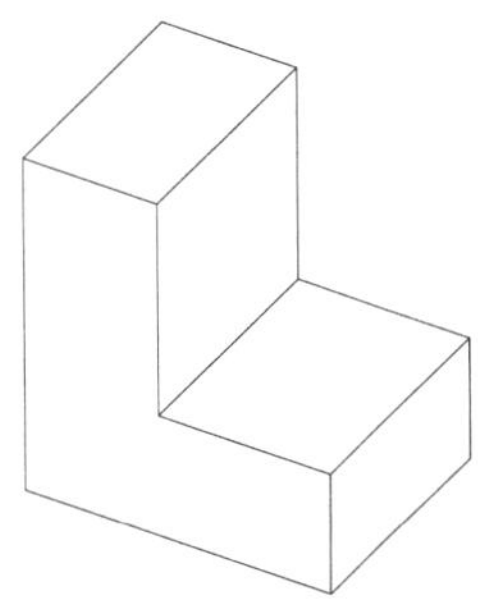

Figure 6–14 Basic L-Shaped Objects

Example 1

How would you go about creating the object shown in Figure 6–16?

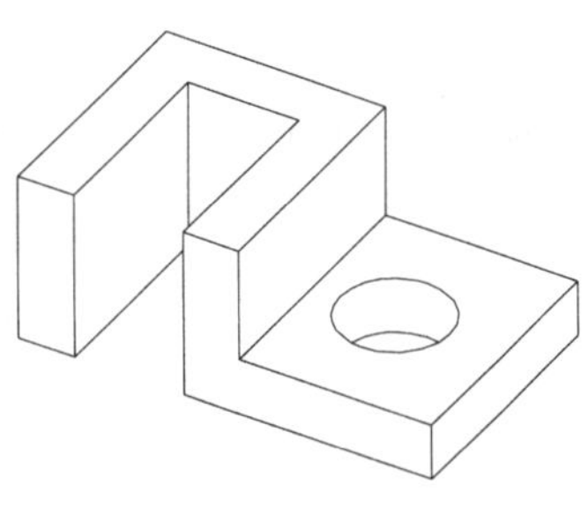

Figure 6–16 Object for Example 1

There are several ways to create the object, however, this example will illustrate the more rigorous method (the given shape can be made with just one extrusion and one cut).

Since the basic shape of this object is a block, start by creating a larger block, then cut a smaller block shape out.

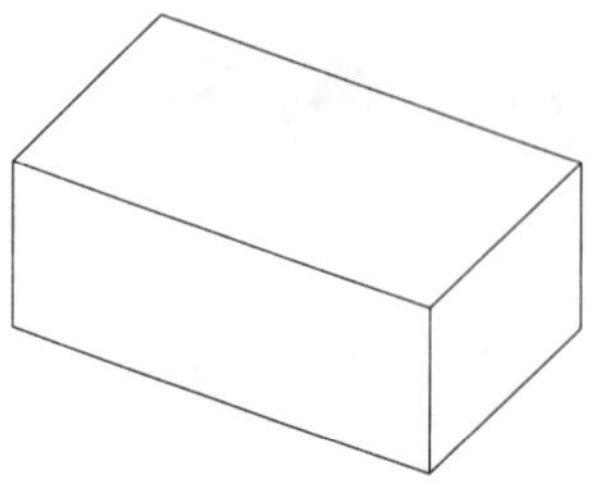

Using the Intent Manager, create a cut similar to the next figure. You could either select the long surface as your sketching surface and cut *Thru All* or select the short surface as your sketching surface and *Blind* to a specific depth.

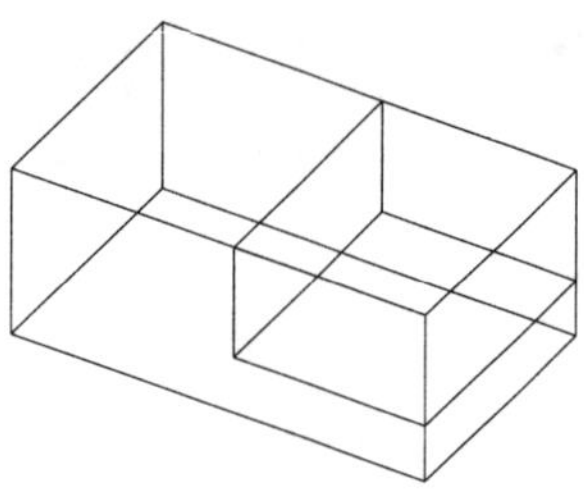

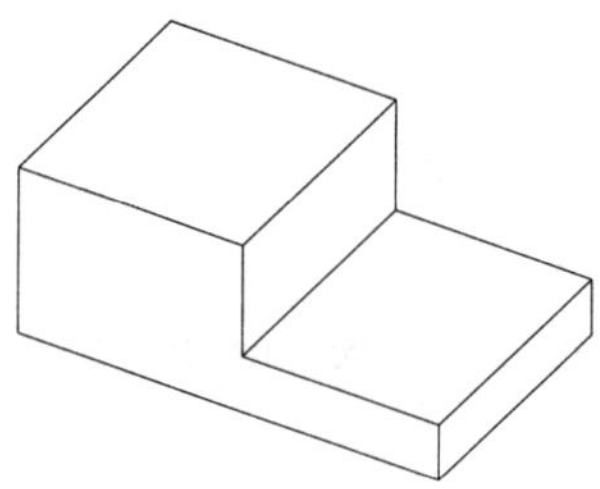

Create another cut, only this time, select the upper surface as your sketching plane.

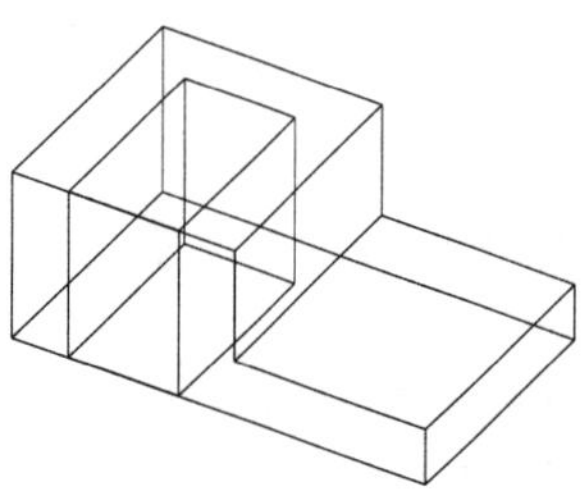

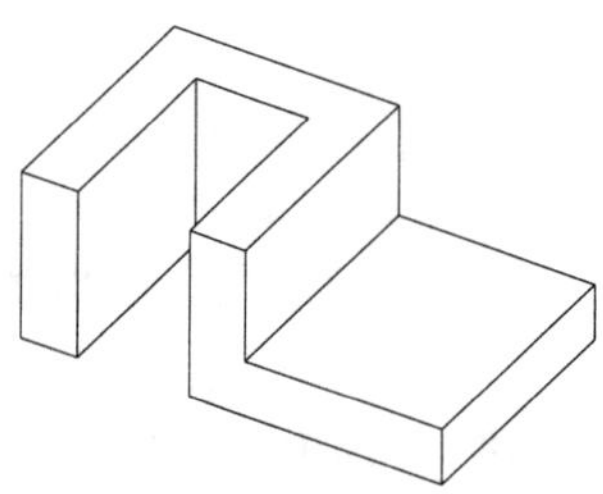

We can finish the shape by creating either a *Hole* or circular *Cut*.

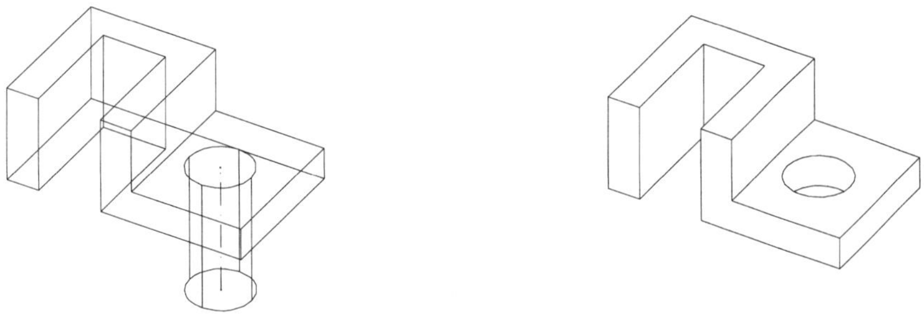

Example 2

How would you go about creating the object shown in Figure 6–17?

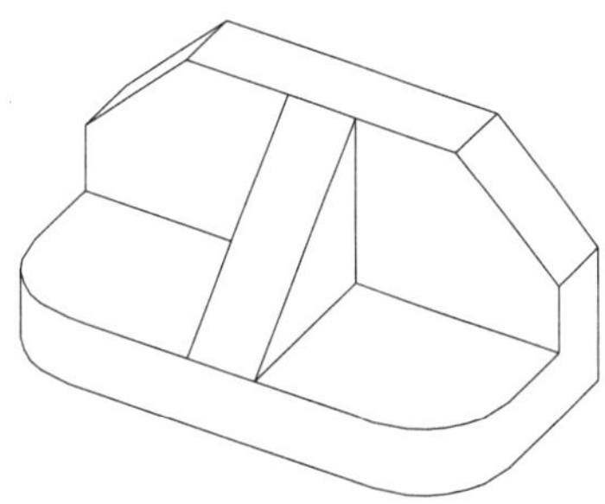

Figure 6–17 Object for Example 2

Again, it would be best if you start by creating a large block and a smaller block to use for cutting.

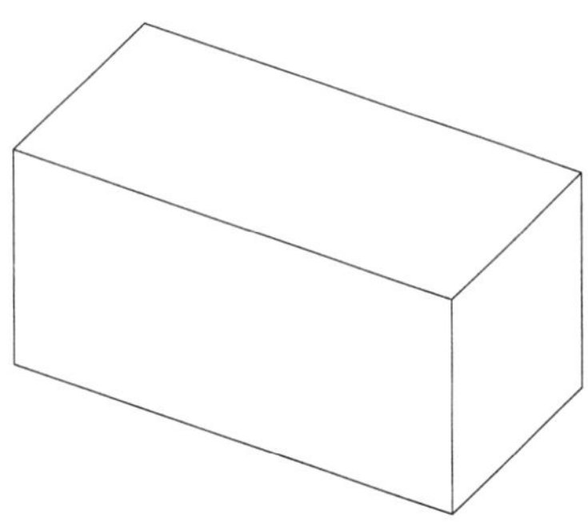

Once the basic block is created, you need to create a *Cut* such that an "L" shape remains.

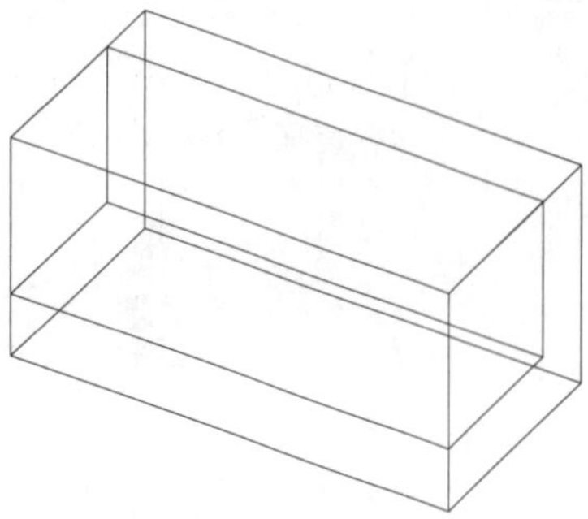

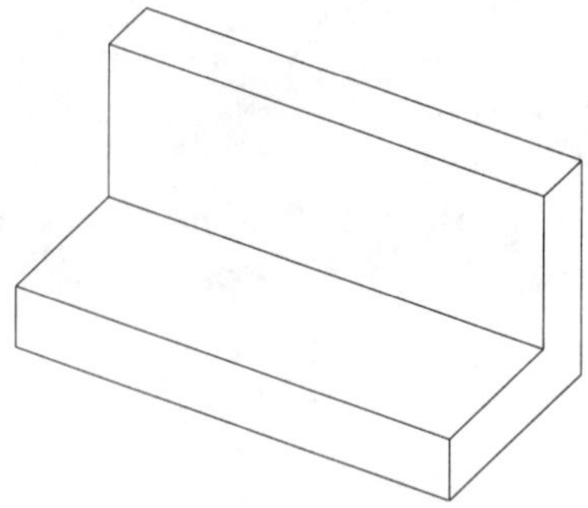

Creating a Rib is a little bit more involved. If you did not do a "two-sided" extrusion, you will have to add a reference data plane. This can be done as you are creating your rib. Select *Feature—Create—Rib—Make* Plane. The *Offset* option is most useful in this case, but you are encouraged to try other options. Once a datum plane has been created in the middle of the "L" extrusion, we simply sketch the shape of the desired rib section and enter its thickness.

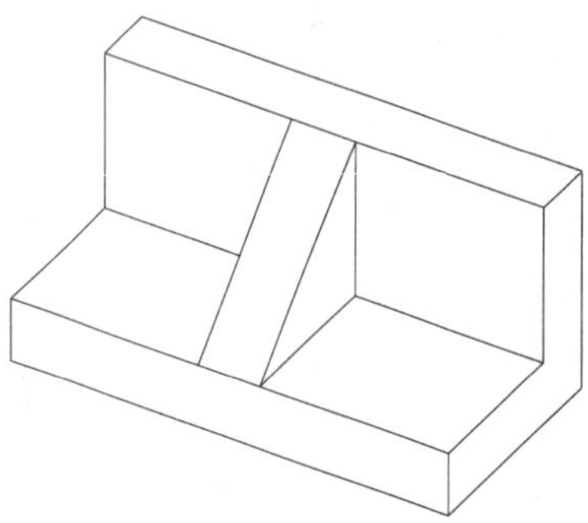

Two options are available to cut the corners, you can either perform a *Cut* operation or use *Chamfer.* Try both methods, one per side.

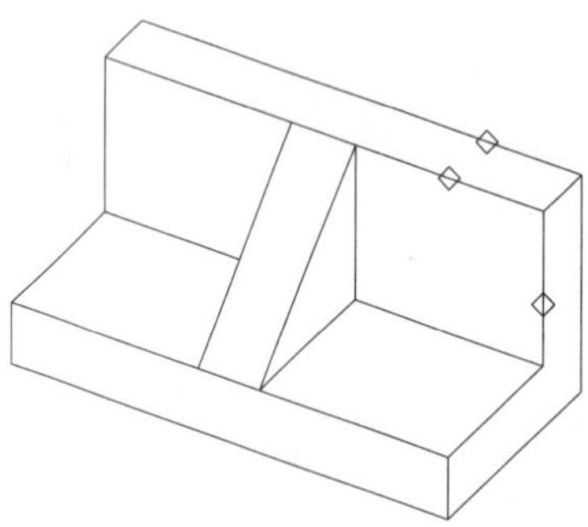

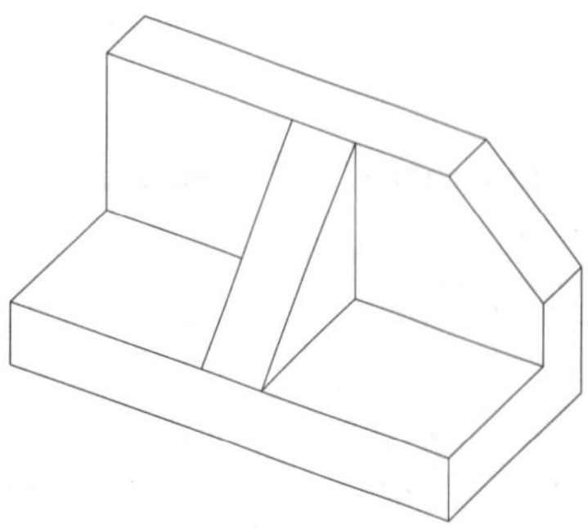

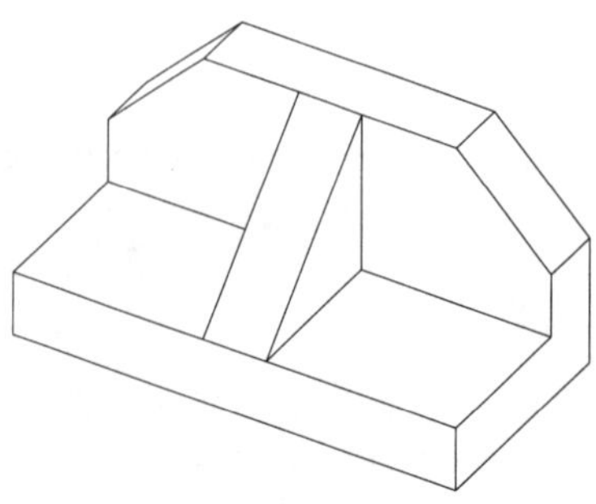

The final feature of this part is created using the *Feature—Create—Rounds* command. When prompted select *Simple—Constant—Edge Chain—One by One*. You will be prompted to select the edges, vertices, or surface to round. Pick the short vertical edges on the base of the part and input the desired radius.

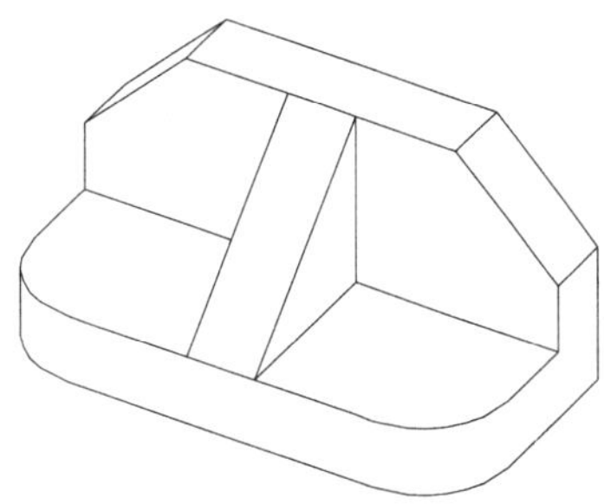

6.2.1 Filleting and Chamfering Strategy

Fillets and chamfers can be added to part geometry in two ways. They can be included as part of your 2-D shape and extruded or revolved to form the feature, or they can be added using the 3-D filleting/chamfering tool. In general, it is easier, and sometimes more reliable, to include fillets and chamfers as part of your sketch before extruding or revolving. However, adding 2-D fillets to a sketch may make it more difficult to modify your object later. (Modifying objects will be covered in Chapter 7.) For the 2-D sketch shown in Figure 6–18a, related dimensions are removed from the sketch when fillets are added as in Figure 6–18b because they were defined between the corner points and the bottom edge. When the corner points were removed, the dimensions went with them.

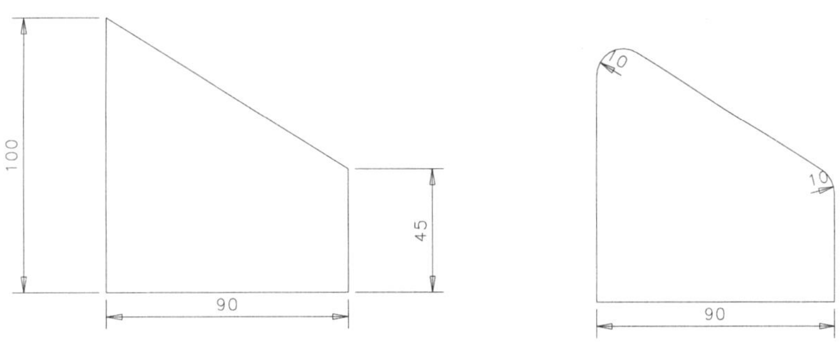

a. Sketch Before Filleting b. Sketch After Filleting

Figure 6–18 Adding 2-D Fillets to a Sketch

If the sketch shown in Figure 6–18b were extruded to form a solid, you would not be able to change the vertical dimensions of the object because there were no dimensions in the 2-D sketch. If the sketch shown in Figure 6-18a were extruded and the indicated 3-D fillets were added to the extruded part afterwards, the vertical dimensions would still be available for modification later. Therefore, adding 3-D fillets instead of 2-D fillets in this situation leads to a more flexible design. As stated previously, when adding 2-D or 3-D fillets and chamfers to a part, it is generally best to add these features last. You may wish to keep an unfilleted version of the sketch or part "just in case" so that you can recover your work if the filleting operation does not work out as you had anticipated.

6.3 Troubleshooting/FAQs

Problem	Potential Solution
You are having problems specifying the three points to create a new datum plane.	Try using *Query Select* and using three surfaces to define your points.
You can't tell which side of the plane you want to keep when performing the Cut operation.	Try using dynamic viewing (*control—middle mouse button*) to see the cutting plane and arrows more clearly.
You are trying to create a 3-D Fillet, but Pro/ENGINEER won't let you do it.	You have probably specified a fillet radius that will not fit in the size provided. For example, if the length of a side is 2, you cannot specify a fillet of radius 5 on the edge bordering the side—it won't fit in the space provided.

6.4 Chapter Summary

In this chapter you first learned about feature-based modeling and some of the standard features available in modeling software. You learned about adding relationships between parts during the construct operation so that your design is flexible and easily modified at a later time. Finally you learned about the usefulness of including reference geometry on your part. Key Words for the chapter are tabulated in the following:

3-D Chamfer	Feature Parameters
3-D Fillet	Features
Chamfers	Fillets
Construct Operations	Point Specification Methods
Coordinate Systems	Reference Geometry
Counterbored Hole	Reference Line
Countersunk Hole	Reference Plane
Cut	Reference Point
Cutting Plane	Relations On/Off
Cylinder	Ribs
Drilled Hole	Slot
Feature-Based Modeling	Sphere

Exercises

On a clean sheet of paper, develop a **Plan of Action** for the object assigned by your instructor, describing the steps you would follow to create the object in Pro/ENGINEER. Use sketches as appropriate in your Plan of Action. Include the features and relations you would apply in creating the object. Then create the object in Pro/ENGINEER, making sure that it is created to the correct *size* as well as the correct *shape*. Object features in the exercises are symmetrical unless otherwise noted. Obtain a view of the object specified by your instructor and turn in a hard copy as appropriate.

1.

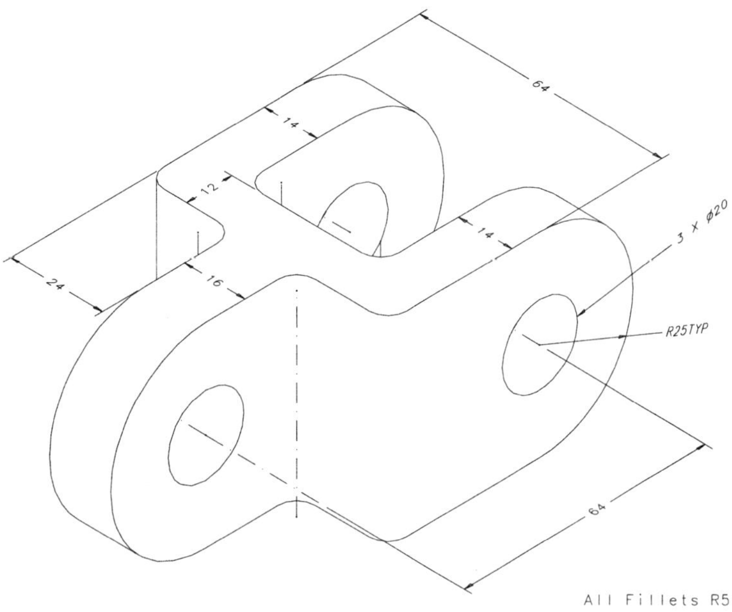
64
14
12
14
3 X Ø20
24
16
R25TYP
64
All Fillets R5

2.

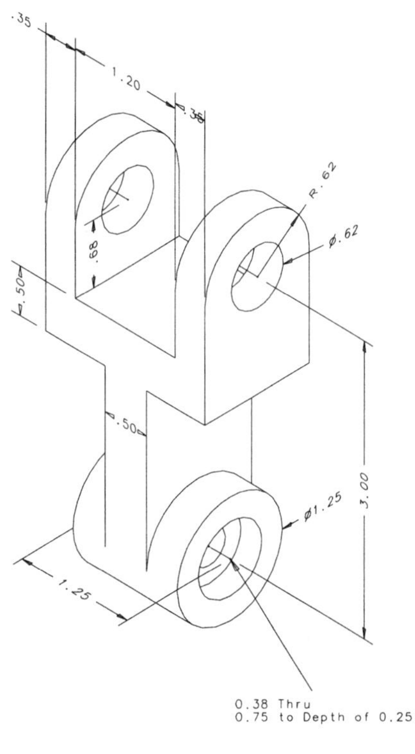
.35
1.20
.35
R.62
Ø.62
.68
.50
.50
3.00
Ø1.25
1.25
0.38 Thru
0.75 to Depth of 0.25

3.

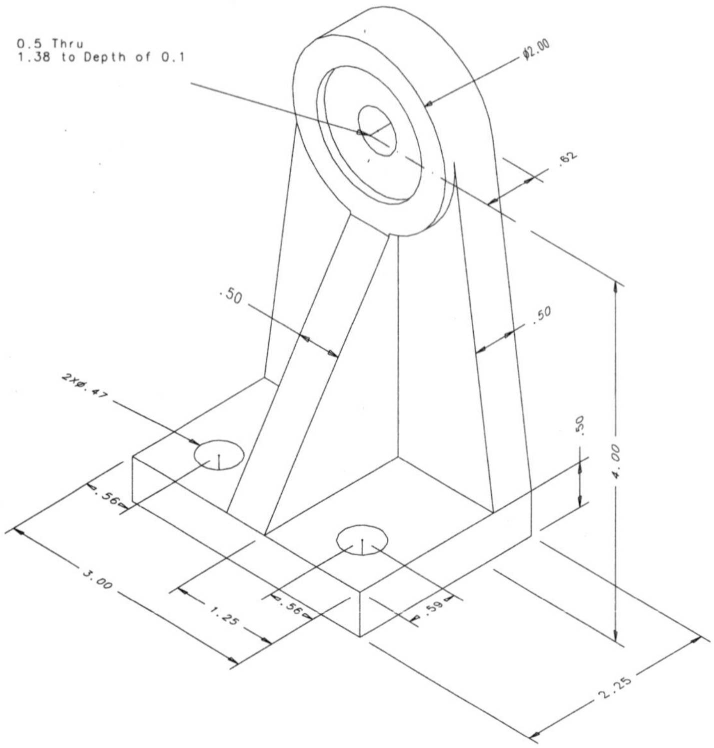
0.5 Thru
1.38 to Depth of 0.1
⌀2.00
.82
.50
.50
2X⌀.47
.50
4.00
.56
3.00
1.25
.56
.59
2.25

4.

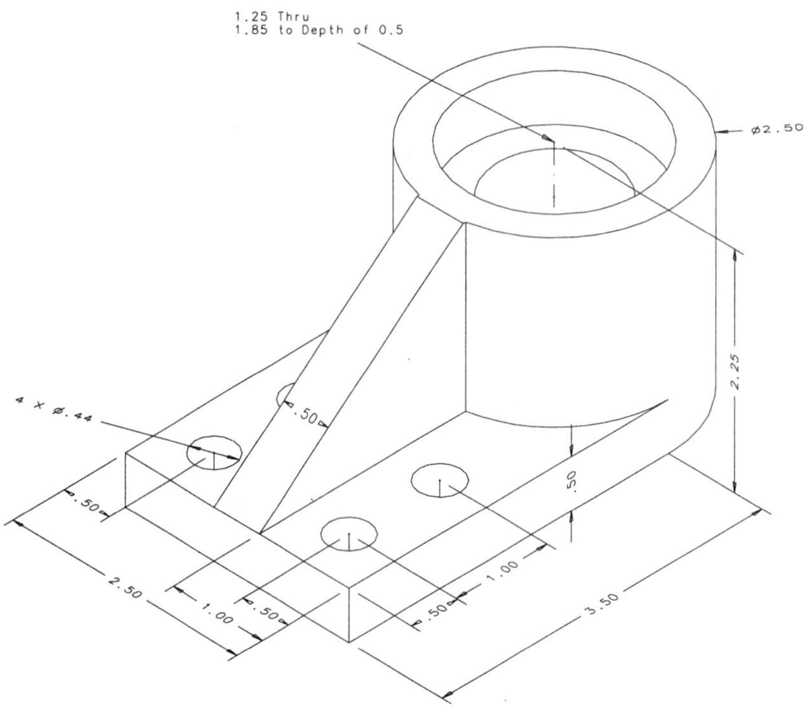
1.25 Thru
1.85 to Depth of 0.5
⌀2.50
2.25
4 X ⌀.44
.50
.50
.50
2.50
1.00
.50
.50
1.00
3.50

5.

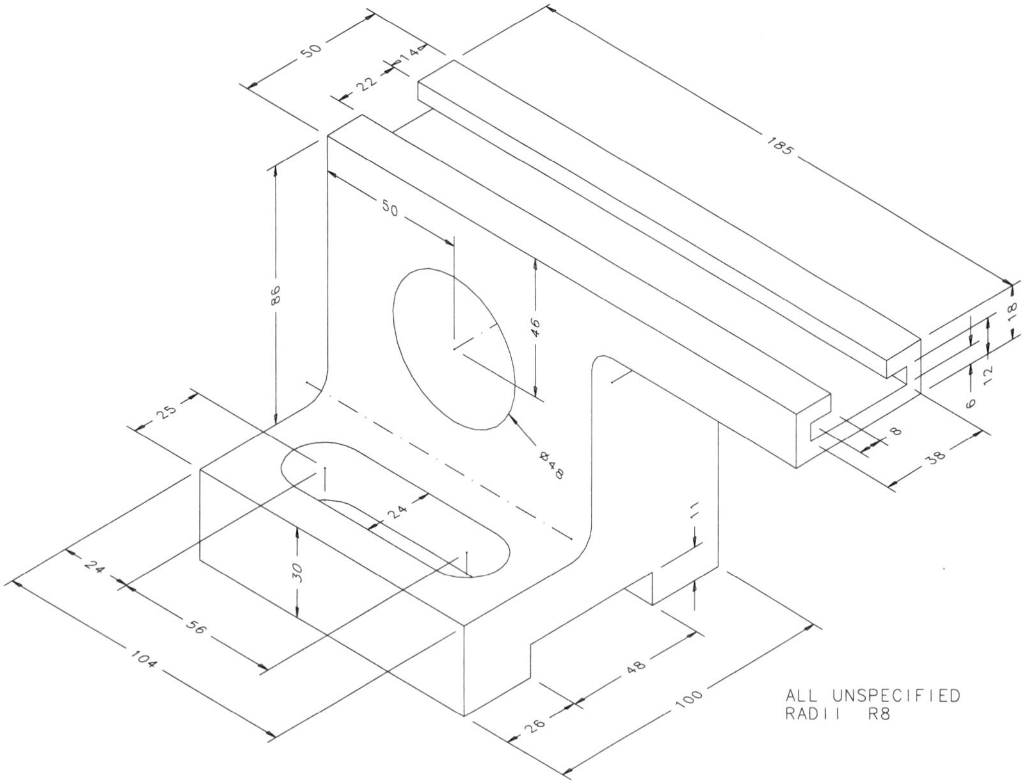
50
22
14
185
50
86
46
18
12
6
8
38
Ø48
25
24
24
30
56
104
11
48
26
100
ALL UNSPECIFIED
RADII R8

6.

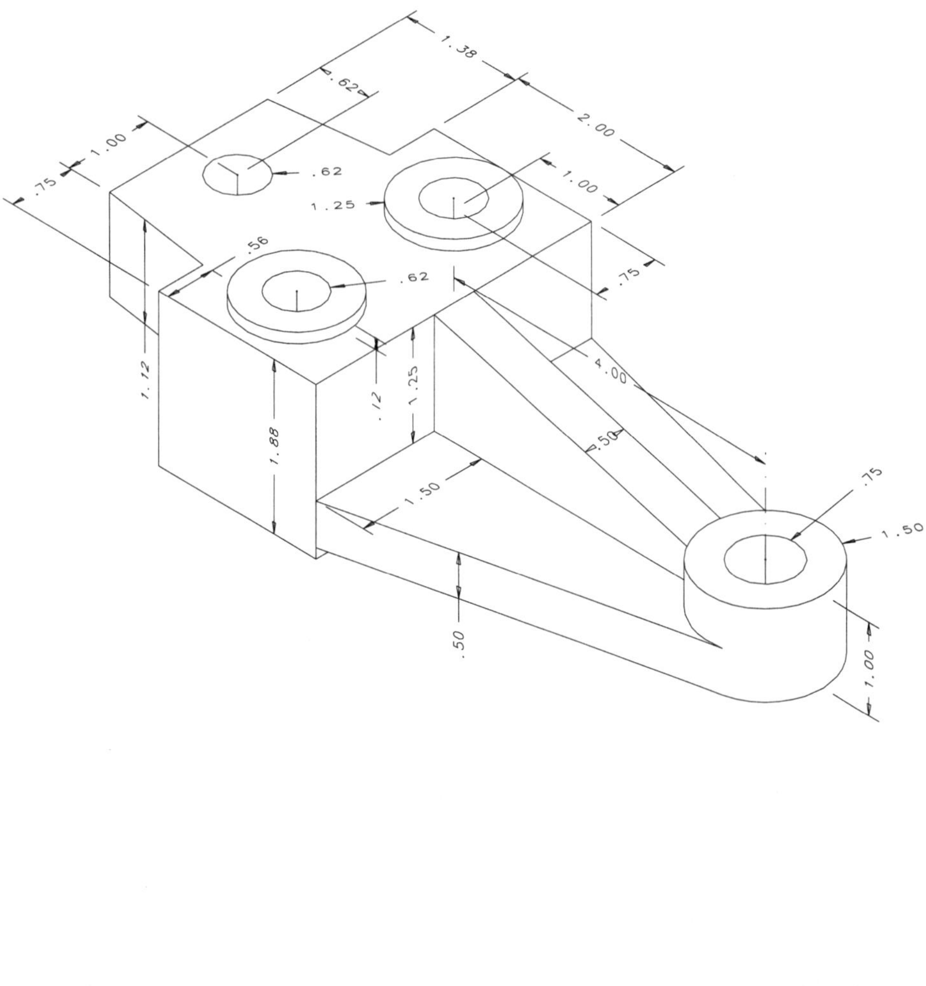
1.38
.62
2.00
1.00
.75
.62
1.25
1.00
.56
.62
.75
1.12
.12
1.25
1.88
4.00
.50
1.50
.75
1.50
.50
1.00

CHAPTER 7

Advanced Modeling Techniques

The most significant advantage that computers have over traditional methods in the engineering design process is the ability to change or modify a design at any time during the design process. First generation versions of most software did not incorporate design flexibility into the solid modeling software. Using these packages, if you decided to change a minor feature of the design, you were often forced to start over and re-create the entire object or assembly. Fortunately, most modern engineering design software has design flexibility built in as one of the primary features of the software. You are now able to easily change objects or assemblies in many different ways, including changing the size and/or shape of the features that make up the individual objects. Features can also be removed from objects as needed to achieve the desired end result. This chapter focuses on how objects are modified using engineering design software. The primary methods for object modification and creation to be discussed are reflecting, copying, modifying individual features, deleting individual features, and advanced features of sweeps and blends.

7.1 Reflecting Objects

Reflection of an object creates a mirror image of the object across a plane. Many times the **plane of reflection** consists of one of the planar surfaces on the original object. In reflecting an object, you will usually indicate whether or not you wish to keep the original half of the object. You could choose to delete the original object and the result will be just the reflected half. Alternatively, you could choose to keep the original half of the object along with the reflected half. This means that your new object will consist of both halves. Figure 7–1a shows an object for reflection. Figure 7–1b shows the result of electing to keep only the reflected half of the original object, and Figure 7–1c shows the result from retaining both halves of the reflected object.

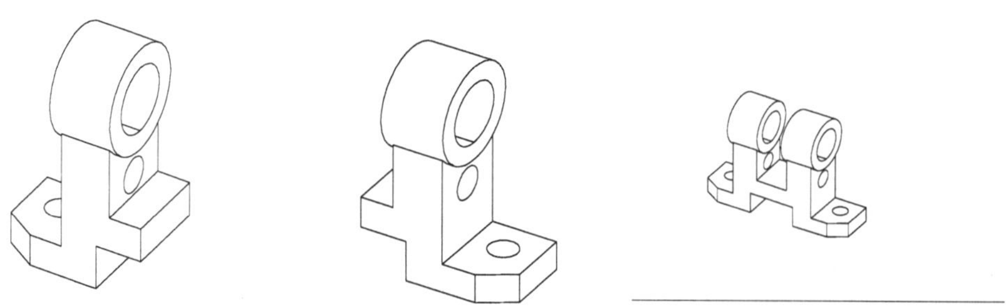

a. Original Object b. Reflected Half c. Both Halves

Figure 7–1 Reflection of an Object

7.1.1 Reflecting Objects in Pro/ENGINEER

The process of reflecting, in Pro/ENGINEER, is called **mirroring**. There are two ways in which Pro/ENGINEER allows you to mirror your part. The first, and easiest, way to mirror the entire part is to select *Feature—Mirror Geometry*. This will create an exact mirrored copy of the original part, which includes all features. Since Pro/ENGINEER reflects in three-dimensional space, you must reflect your geometry about a plane. This may be unlike what you have used in two-dimensional CAD packages, where you reflected geometry about a line on the page. Typically, you will be reflecting a part about one of the surfaces on the part itself. After picking the plane of reflection, you will see the new geometry automatically. An alternative to selecting a plane on the original geometry is to create a new datum. This is done by selecting *Feature—Mirror Geometry—Make Datum*. This method allows for more advance mirroring, such as offset mirroring or angled mirroring.

The second, and more intricate method used to mirror your part is done by selecting *Feature—Copy—Mirror*. This is a somewhat more advanced approach, as it allows you to mirror some or all of the features on your original part. (Recall that every cut or protrusion is a separate feature of your part.) When mirroring your part in this manner, careful attention must be given to ensure that only the selected features are mirrored. With a complex part, it is often difficult to select all the features of your part. In this case, you should use the *Feature—Mirror Geometry* command.

Guided Tour 7.1

- Start up Pro/ENGINEER, and create a new part. Specify millimeters as your units. Create a block with the following dimensions: length = 30, height = 50, and depth = 50. Create a lug by specifying the following parameters: width = 30, total length = 40, hole diameter = 20, and lug thickness = 10. Create the lug such that it is in the lower corner of the block, and an object similar to the next figure results.

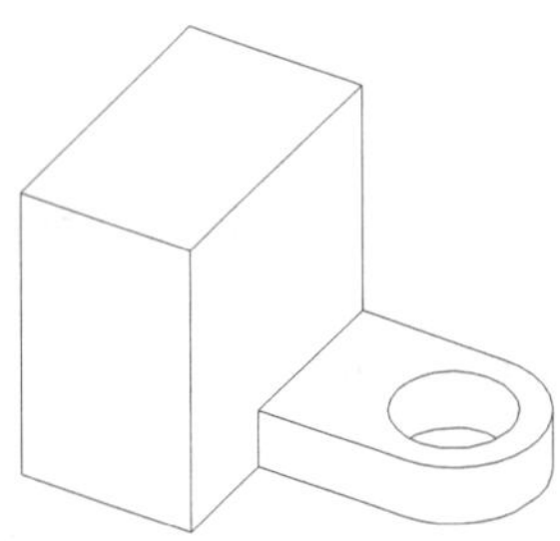

- Make sure that you have a wireframe view, then reflect the part about the vertical plane defined by the side opposite the plane that the lug is attached to. When finished, your object should look like the following:

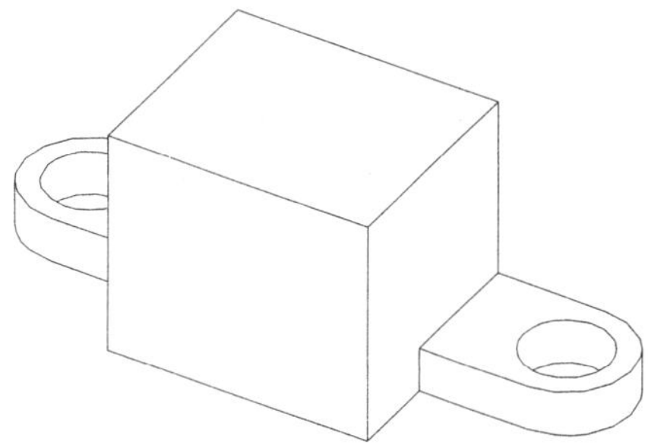

7.2 Copying Objects

When working with some solid modeling software packages it is often necessary to orient or move objects in space. Not all solid modeling software requires this for efficient operation, but most do. Two general ways in which an object can be moved in space are by *translation* or by *rotation*. Often, you have the ability to make copies of the original object as you are performing the orientation command. With this feature, you can translate (or rotate) *and* copy the object at the same time. This means that the original object is left in place and copies of it are placed in the new positions. The commands to create copies of the original part are *Feature—Copy—Move*. This allows for a copy and translation of any or all features of the part. Please see Figure 7-2.

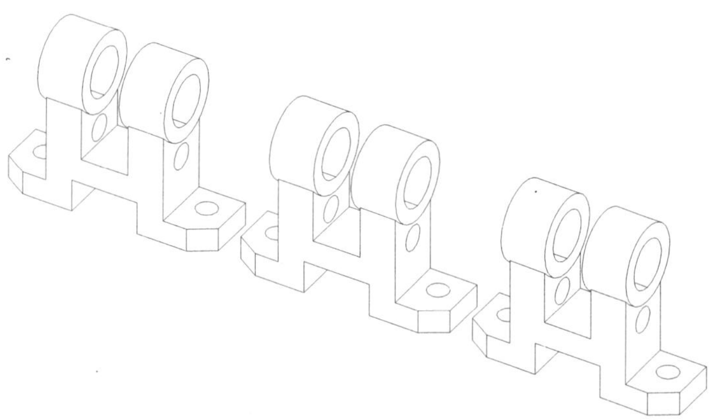

Figure 7–2 Object Translated and Copied

7.3 Modifying and Deleting Individual Features in Pro/ENGINEER

A useful tool for part development and modification in Pro/ENGINEER is the part Model Tree. The Model Tree is a window that keeps track of the features of the part you are building. The Model Tree also shows the part's history, as earlier features are shown on the top and later features near the bottom. The Model Tree is very useful when you wish to modify your part. For example, it is often difficult to select a particular feature to delete from the part geometry. An easier way is to select *Delete* from the Menu Manager and then select the feature from the Model Tree.

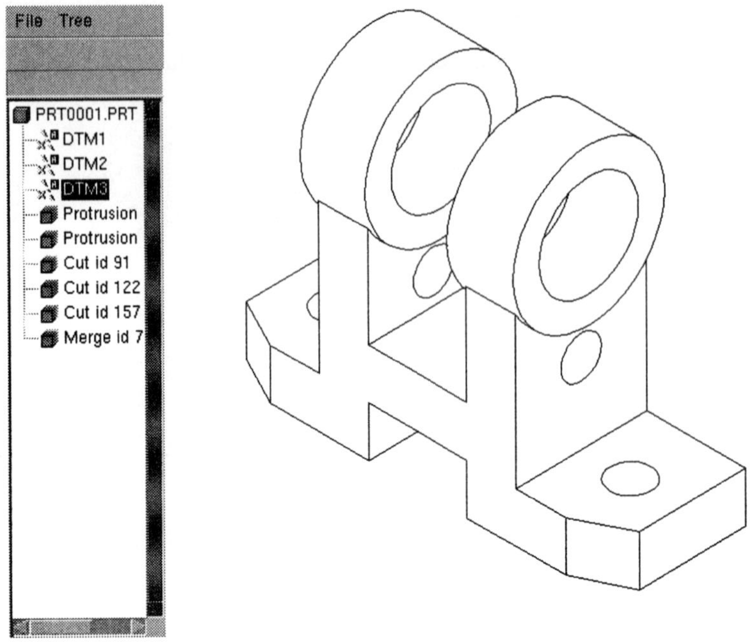

Model Tree for the Part Shown

7.4 Cross Sections and Paths

You have learned about the basic methods for creating objects—extrusion, revolution, combining objects, and adding basic features. There are many objects that can be created by these methods, however, there are fairly complex objects that could not be modeled by these methods such as automobile bodies, airplane wings, boat hulls, propeller blades, and many others. Two techniques commonly available in solid modeling software that enable you to create these complex shapes are *sweeping* and *blending*. The remainder of this chapter will focus on these two methods and on how they can be used to create complex three-dimensional geometric shapes.

Both the blending and sweeping operations require that you define **cross sections** and **paths** for the part that you are creating. A cross section is a two-dimensional shape that defines what the object would look like if you sliced through it. You are already familiar with the concept of a cross section from your experience with extrusion. Extrusion produces an object of constant cross section—no matter where you slice through the object, you will always obtain a cross section that is the identical to the 2-D section that you extruded (provided that your slice is taken perpendicular to the axis of extrusion). Figure 7–3 shows an object and three different cross sections that are obtained by slicing through the object with the indicated cutting planes.

A path is used primarily in a sweeping operation to orient the cross sections in space. With sweeps, you can have an object that is created by locating cross sections along a curved path in space. As the cross section changes from one location on the path to another, the software figures out how to transition the object surfaces from one cross section to another. In the extrusion process, the path for an object created by extrusion is a straight line—the axis of extrusion.

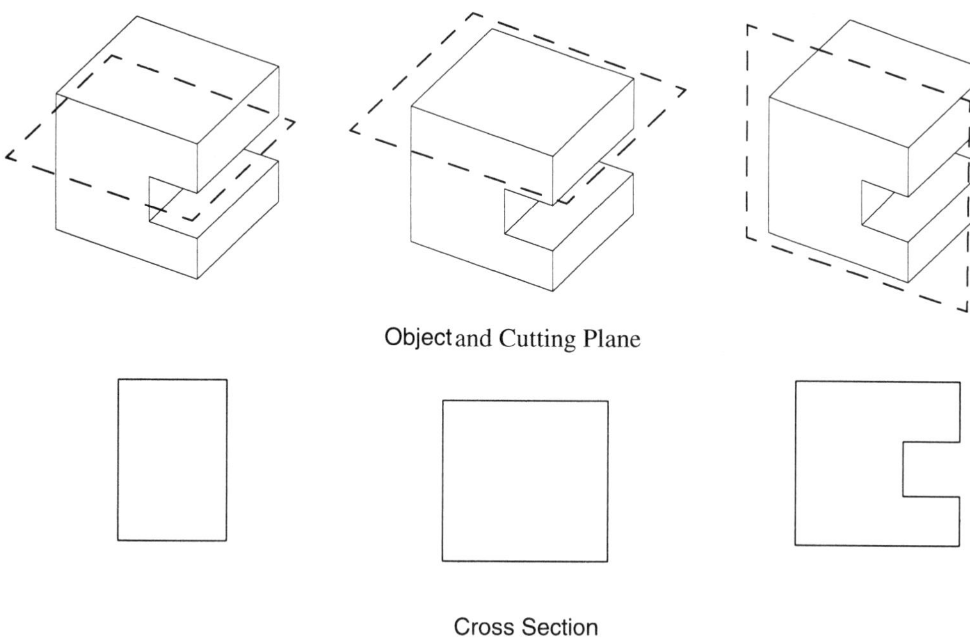

Figure 7–3 Various Cross Sections of an Object

Blending is accomplished by orienting varying cross sections in space, and sweeping is used for objects that have a curved path and either constant or varying cross sections. The first object in Figure 7–4 was created by varying the cross section from one location to another (blending), and the second object was created by maintaining a constant cross section along a curved path (sweeping).

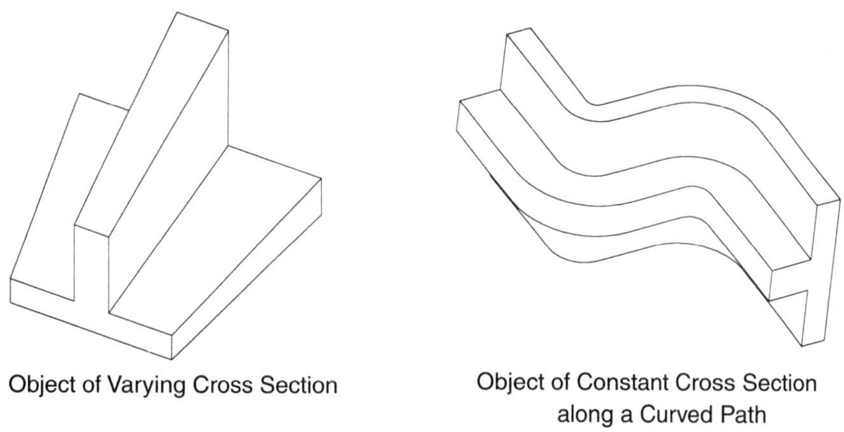

Figure 7–4 Objects Created by Blending and Sweeping

To create blended or swept parts, you will draw or sketch the path curve (if you are sweeping), draw or sketch the 2-D cross section(s) required to create the part, and orient and constrain the cross sections in space (if you are blending).

7.5 Creating Objects by Blending

Blending is typically used to create objects of varying cross section, however, it could also be used to create objects of constant cross section where the cross sections are oriented in space along something other than a straight-line path. (An object that has a constant cross section along a straight-line path is best created by extrusion.) Figure 7–5 shows the initial cross sections and the final objects for two objects created by blending. In the first object, the cross section was varied from one location to the other, and the cross sections were located along a straight-line path. For the second object, the cross section was held constant, but copies of the cross section were oriented in space along something other than a straight-line path.

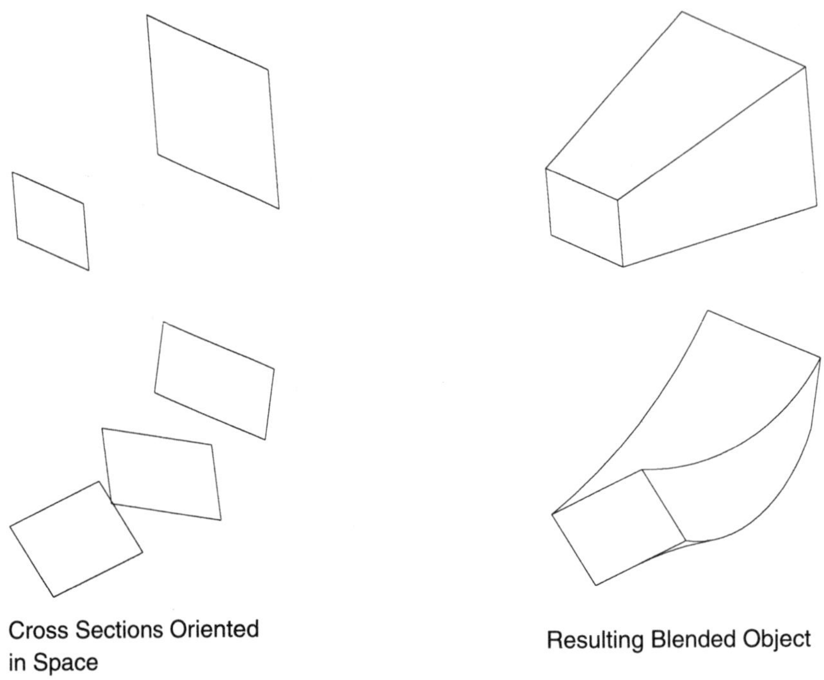

Figure 7–5 Objects Created by Blending

When creating blended objects, you must be careful to make sure that you don't inadvertently create "twisted" parts. Twisted parts will sometimes occur depending on *how* the cross sections were selected for inclusion in the blended part. When selecting sections for a blended part, you must cursor pick a "similar" location on each of the sections that make up the part. The software creates blended parts by mapping one corner point (usually the one closest to the point where you selected the first section) to the first point on the next section, and so on. It will then proceed around each section (either counterclockwise or clockwise depending on the software) mapping the next corner point on the first section to the next one on the second section. For this reason, if you select the sections at dissimilar locations, you may end up with a twisted part. Figure 7–6a shows two sections oriented in space. Figure 7–6b shows the resulting blended part that occurs when the first section was cursor selected for blending near location A on the section and when the second section was cursor selected near location B on the section. Figure 7–6c shows the resulting blended part that occurs when both sections were cursor selected for blending near location A on each section. Note the twisting of the part when the sections were selected at dissimilar locations on the perimeter.

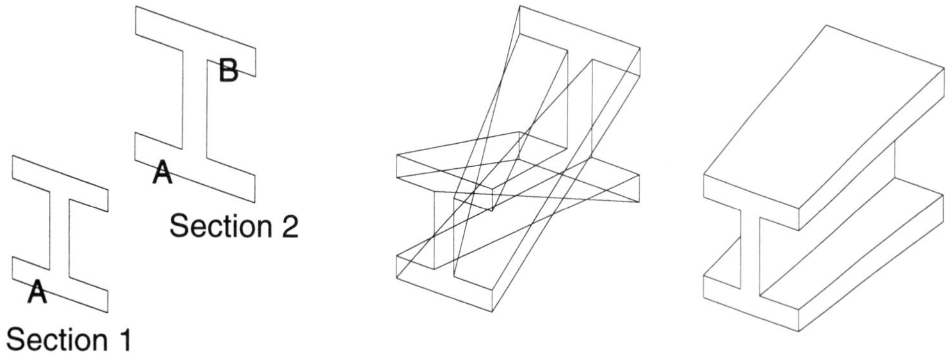

a. b. c.

Figure 7–6 Twisted and Non-Twisted Blended Parts

Blending is generally used to create solid objects by including closed cross sections in the part definition. However, in some instances, you may wish to create a thin surface in space by blending between open curves. The procedure you follow is similar to creating an object from closed sections. Figure 7–7 shows two open curves oriented in space and the resulting blended open surface as a shaded image that is obtained from these curves.

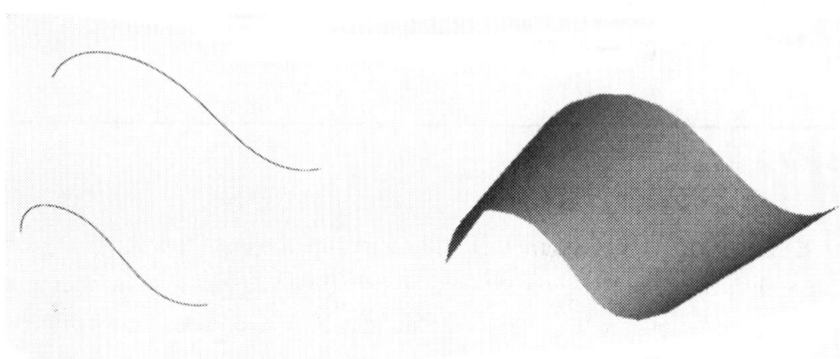

Figure 7–7 Blending Open Surfaces

Another feature of most blending commands is the ability to set whether the finished part should be "smooth" or "straight." The smooth option maps the transition from one section to the next with smooth, spline-type surfaces. The straight option creates linear interpolated surfaces between the sections. Figure 7–8 shows the difference between the smooth and straight options, given the same three cross sections.

7.5.1 Blending Solid Objects in Pro/ENGINEER

The *blend* option in Pro/ENGINEER is found at the same location as the *Extrude* and *Revolve* functions. To create an object by blending in Pro/ENGINEER, you select *Feature—Create—Protrusion—Blend—Solid—Done*. This will require you to create a closed section geometry. (If you wish to create an open surface, select *Thin* instead of *Solid*.) For ease of explanation, we are going to stick with the *Parallel—Regular Sec—Sketch Sec.* options. After you feel comfortable with how the blending options work, you are encouraged to experiment with the other options. The blends you will create can either be *Straight* or *Smooth*.

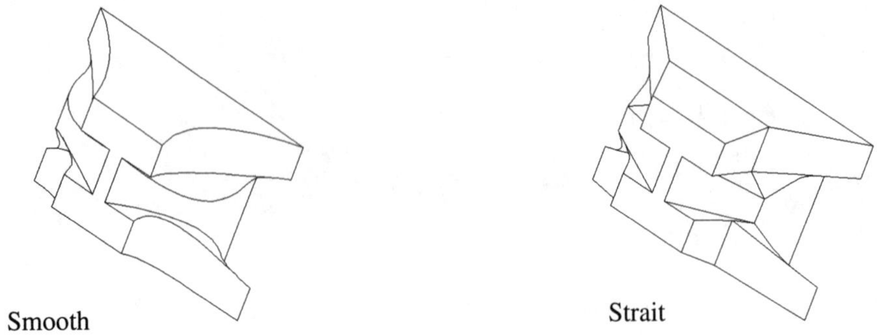

Figure 7–8 Blending Options

Once the references are defined (you are using datum planes and the intent manager right?) you can begin to sketch the cross sections. When you are satisfied with the location, size, shape, and constraints of the cross section choose Section Tools (*Sec. Tools*) and select *Toggle. Toggle* allows you to sketch and constrain the next cross section and return back to earlier sections if editing is required. Once you are satisfied with all the cross sections, select Done. You will then need to enter the distance between each section. Select *Preview* and use dynamic viewing to see how your blend looks. If the preview part is acceptable, click *OK*, and the blended solid will be created. If you have a problem with your blended part, such as twisting, you can do one of two things. Either click *Cancel* to start over in defining the blended part, or redefine any of the elements associated with the blend by highlighting the element in the Element Info box and clicking *Define*. If you *Cancel* the blend operation, you can reselect it, taking more care in how you cursor select the sections for blending. If you choose *Define*, be sure to select the appropriate element which you wish to edit.

Guided Tour 7.2

- Start up Pro/ENGINEER and make sure that millimeters are the units specified for your new part. Beginning in the upper left corner of the geometry and proceeding in a clockwise manner around the shape, sketch the cross section shown in the following. You may have to add or remove dimensional constraints so that your final dimensions are specified as are those shown in this figure. Note that the geometric constraints are not shown in this figure, for clarity's sake.

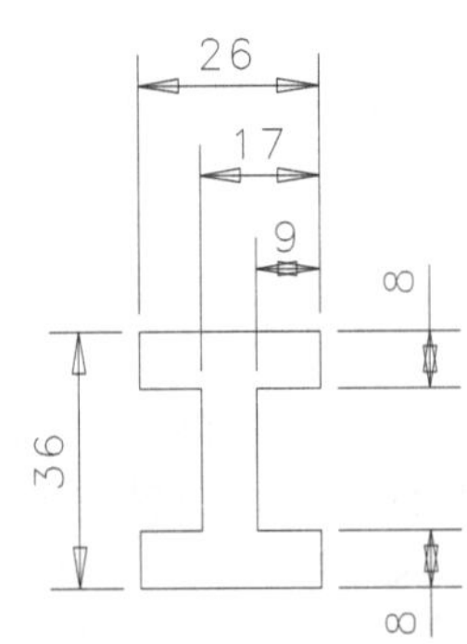

- Select *Section Tools* and click on *Toggle*. The previous section should now appear grey. Start a second section, but change the dimension on the section that is currently 36 to 20 instead.

- When finished with the second section, select *Done* and enter a depth of 65 mm.

- Preview the part to make sure it appears correct and then click *OK*. Your part should look similar to the following one:

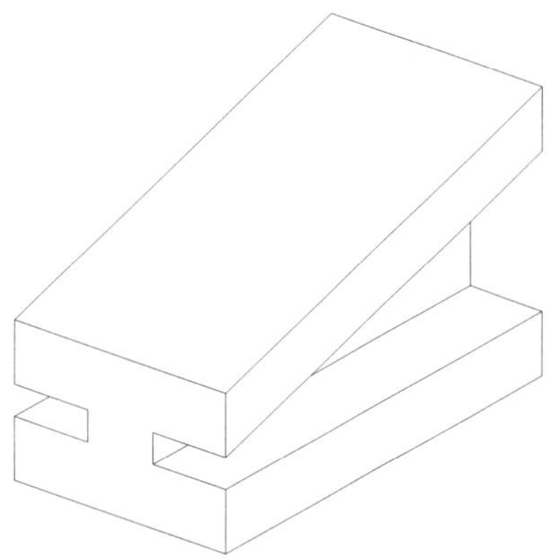

7.6 Creating Objects by Sweeping

Swept parts are created by "sweeping" cross sections along a curved path. In most cases, you will sweep a constant cross section along the path. The path that you choose to sweep along can be either open or closed, depending on your needs. Figure 7–9 shows two objects. One object was created by sweeping a circular cross section along a closed path, and the other object was created by sweeping the same cross section along an open path.

Circular Section Swept Along a Closed Path

Circular Section Swept Along an Open Path

Figure 7–9 Swept Parts of Constant Circular Cross Section

Sweeping can also be accomplished by specifying more than one different cross section along the path. In this way, you can obtain a swept part that tapers from one cross section to another along the path. Figure 7–10 shows a swept part that was created by transitioning between cross sections of two different sizes along the curved path. Note that when you create parts by this method, you specify the size at each end of the path, and the software transitions between these sections along the path. If you specify three sections along the path, the first section will be oriented at the beginning of the path, the second section will be located at the middle of the path, and the third section will be located at the end of the path with transitions in between.

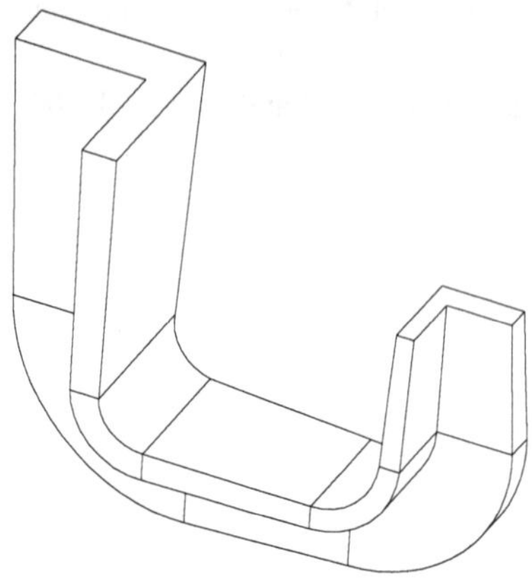

Figure 7–10 Swept Part of Varying Cross Section

When performing sweep operations, you can avoid problems in your modeling by carefully considering three characteristics of the object you will sweep. Two considerations involve the corners of your path—the sharpness and the fillet radius. The third consideration is the point of the orientation for the section.

The path for the sweep operation cannot contain any sharp corners or discontinuities in it. Because the sweep is performed as a continuous operation, the software would not be able to transition around a sharp corner. This does not present an insurmountable problem because fillets can be added at sharp corners on the path.

If fillets are included at corners in the path profile, you must take care to specify them of a sufficiently large radius so that the section does not interfere with itself as it is swept around the filleted corner. The size of the required fillet radius will depend on the size of the cross section used to define the part. Figure 7–11a shows the result of sweeping a cross section around a filleted corner specified by a relatively small fillet radius, and Figure 7–11b shows the part that results when the fillet radius on the path curve is increased so that it is more proportional to the size of the cross section. Notice that for the object created with a small fillet radius, the interior portion of the section is incorrectly modeled at the corner location. Situations such as these should be avoided in creating solid parts by sweeping.

The final thing you must consider when creating parts by sweeping is the orientation of the cross sections with respect to the path. In general you will be creating the two-dimensional cross sections and the curved path all on the same plane in space. Thus, when you are selecting the cross sections and paths to use in the sweep operation, you will need to be able to locate them relative to one another. The way that you locate the sections on the path will vary among software packages. Typically cross sections will be oriented along the path according to their centroids. In other words, the cross sections will be oriented so that the path curve goes through the center of gravity for the section. Sometimes, however, you may wish to orient the section so that the path goes through a different point. Frequently this other point is a corner point on the section itself. You will usually obtain different results from the sweeping operation depending on how you orient the sections relative to the path. Figure 7–12 shows two objects created by sweeping cross sections of varying sizes along a curved path. For the first object, the cross section was oriented by its center of gravity along the path and the second object was obtained by orienting relative to a corner of the section. Notice that the upright sides of the first object are

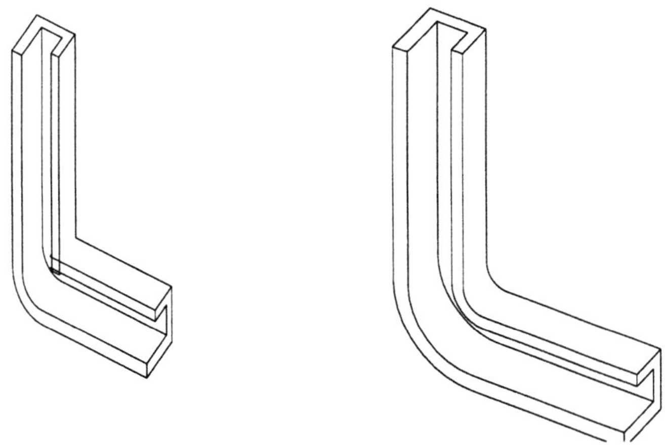

a. b.

Figure 7–11 Sweeping a Cross Section Around a Filleted Corner

flared in both directions, whereas the upright side of the second object is only flared out to the left but is vertical on the right side.

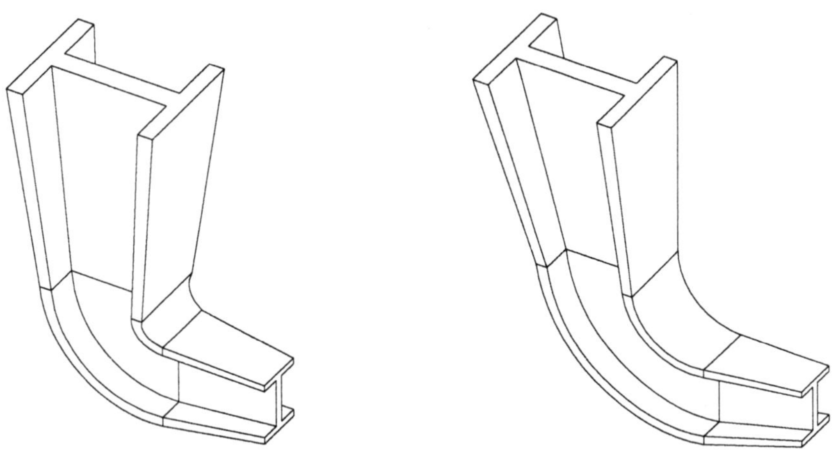

Figure 7–12 Orienting Cross Sections Through Different Points

In addition to orienting the sections relative to a point along the path, you might also need to specify the rotation of the section about the axis of the path. In general, you will obtain very different objects by rotating the sections differently along the path. Figure 7–13 shows two objects created by sweeping an identical section along an identical path. The difference between the two resulting objects was that the section was rotated by 90 degrees for the object on the right.

7.6.1 Sweeping Solid Objects in Pro/ENGINEER

Creating an object by sweeping in Pro/ENGINEER is a two-step process. You must first draw the path for sweeping (it can be either an open or a closed path) and then draw the cross section to sweep. One advantage to sweeps in Pro/ENGINEER over other software is that Pro/ENGINEER will allow sharp corners in your path definition. You can include fillets or arcs at corners in order to create a smooth sweep. The cross sections will automatically be located perpendicular to the path during the sweeping operation, so you can initially sketch all of the 2-D geometry on one reference plane.

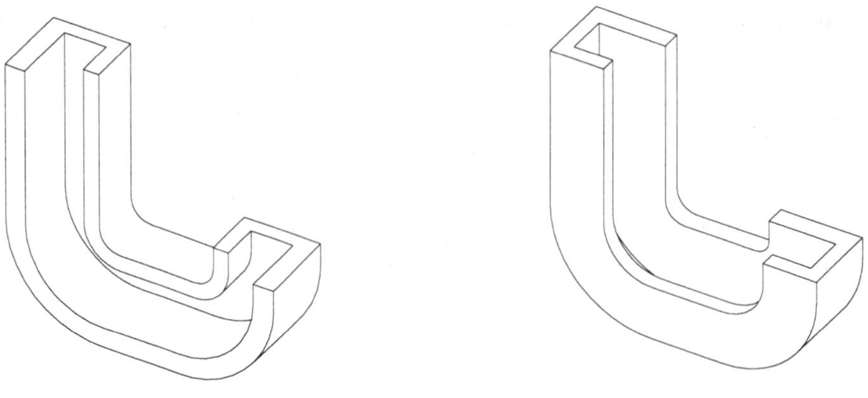

Figure 7–13 Rotating Cross Sections Along the Path

Pro/ENGINEER, also allows you to create open and closed sweep paths. If you wish to create an open path sweep, it is important to remember that your cross section must be closed. However, for closed paths you have the option of creating internal faces. An internal face allows you to have a solid object when finished. Figure 7-14 shows the difference between a part with and without an internal face.

Without Internal Face With internal Face

Figure 7–14 Closed Sweep Paths Options

When sketching the path, an arrow will appear to show the direction of the sweep. Once you are satisfied with your path shape, select *Done,* and Pro/ENGINEER will automatically align the sketching plane perpendicular to the sweep path. If you are unsure of the alignment, you can use the dynamic viewing controls to see the relationships. After using the dynamic viewing options, be sure to select *Sketch View* from the Menu Manager. A typical sweep path is shown in Figure 7-15.

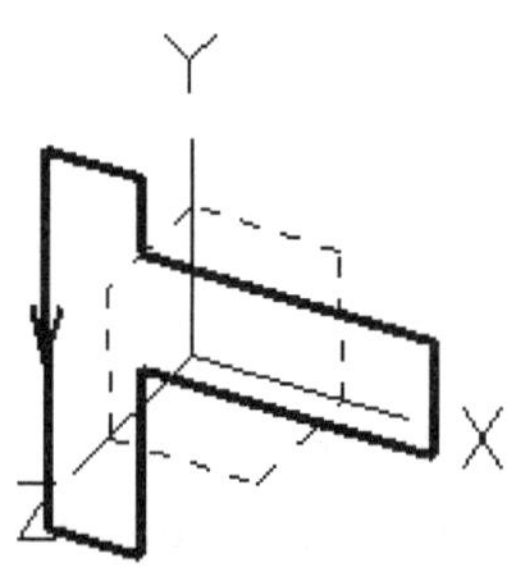

Figure 7–15 Typical Sweep Path

You should preview the part by clicking the appropriate button on the form and using dynamic viewing to inspect the part. If the part looks as it should, click *OK*, and the solid object will be created. If after previewing the part, it is not what you want, you can redefine the cross section or paths by clicking on the *Define* button.

Guided Tour 7.3

- Start a new part labeled *Bracket*. The creation of this part will use protrusions, blends, sweeps, and cuts. We will start the construction without first creating datum planes because we will add them later. Make sure to change the unit system to millimeters. A preview of your final shape is shown in the following figure:

- First create a base which is 70 mm x 40 mm by 15 mm thick.

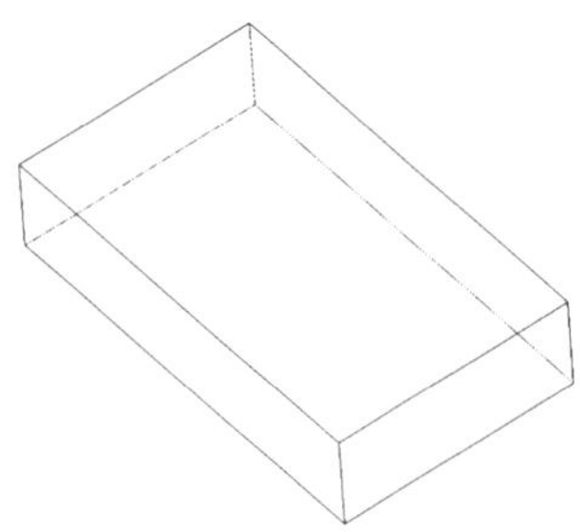

• Create a straight blend with cross sections centered on the base.

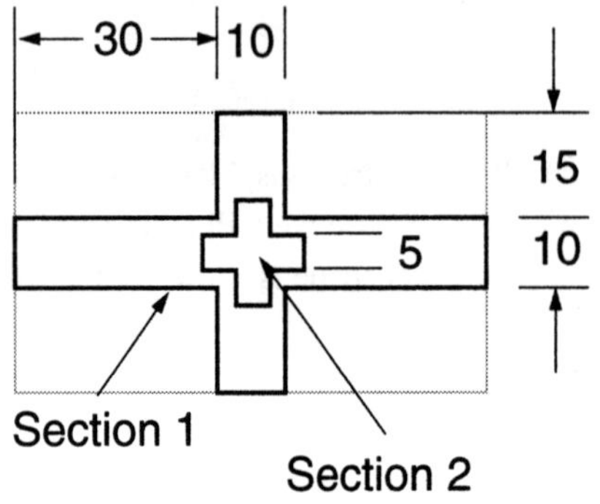

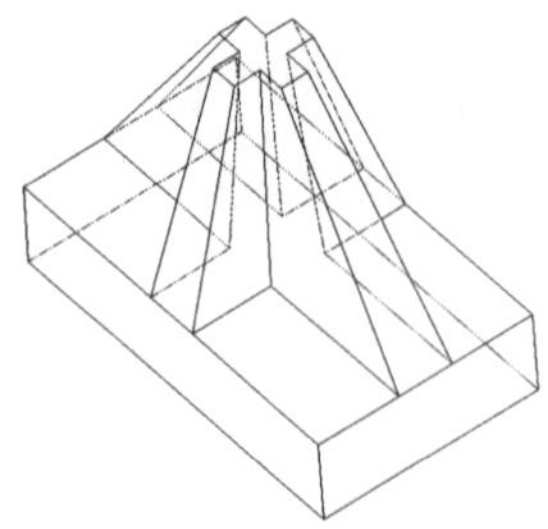

• Before we can create a sweep, we must create a datum plane. Select *Feature—Create—Datum—Offset.* Select one of the 70 mm x 15 mm surfaces on the base. The offset distance is 20 mm. This should place a datum plane bisecting the part.

• Create a sweep using the same cross section as Section 2 from the preceding figure. Choose the newly created datum plane as the sketching plane for the sweep path. The sweep path is to consist of a 25 mm vertical line centered on top of Section 2, followed by a half circle 35 mm in diameter.

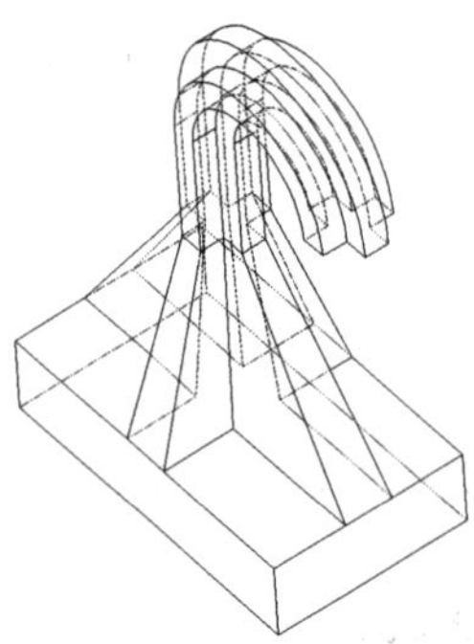

- Add a cylinder with a diameter of 20 mm. Align the center line of the circle to the end of the swept section. In the middle of the cross section, *Protrude* this to a depth of 25 mm (Hint select the datum plane you created earlier as your sketching plane, and do a two-sided protrusion).

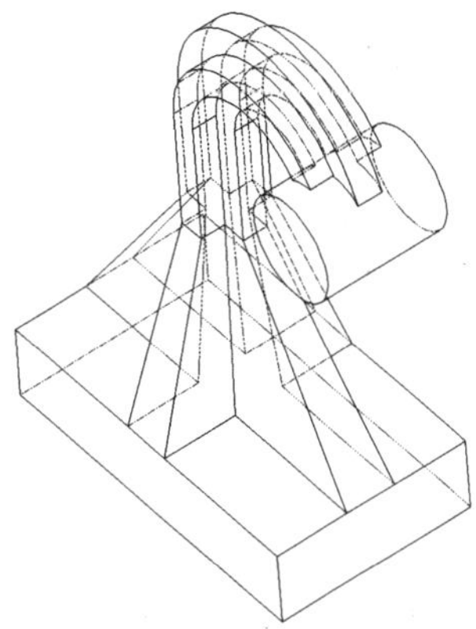

- Finally, cut a 10 mm diameter hole through the cylinder.

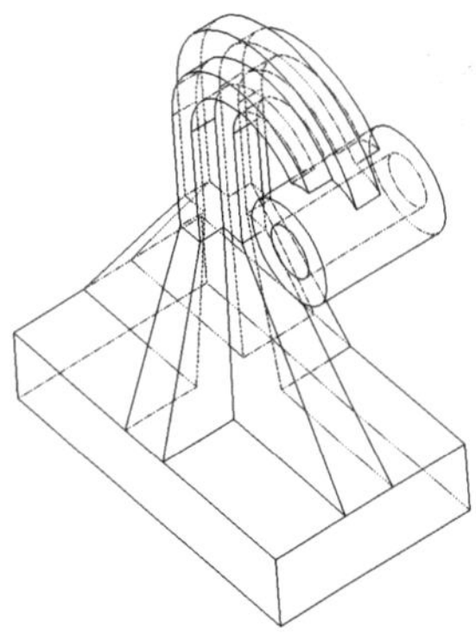

7.7 Blending and Sweeping Strategy

The most important thing to remember about blending and sweeping is to always preview your part geometry before you create it. It is much easier to fix things up front, when it is still easy to redefine and make modifications to the swept or blended feature. The other thing to remember about blending and sweeping is that these operations are really a "last resort." Blending and sweeping should only be used if other methods are not practical for the creation of an object. The reason for this is that objects created by other means (extrusion, revolution, cut, etc.) are generally easier to modify than are those created by blending and sweeping. In the following examples, objects are shown that could be either blended and swept or created by another means. The method for creation is described for each.

Example 1

The desired part is the tapered circular object shown in Figure 7–16. What are the two alternate methods to be employed in creating this object?

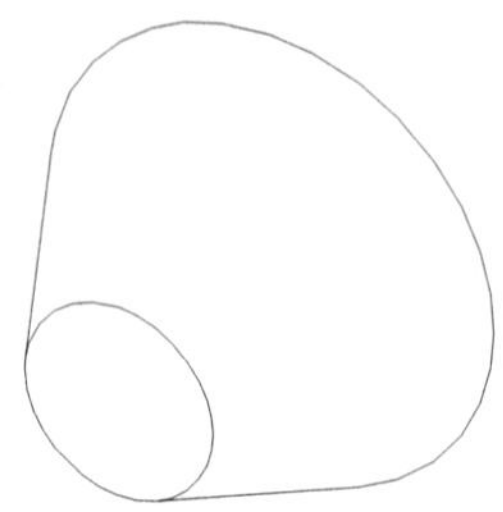

Figure 7–16 Object for Example 1

To create this object by blending, you would first create two circles of the appropriate sizes, then specify the depth, Figure 7–17. Another approach is to create this shape using the *Revolve* option. This simply entails the sketch of an edge and a center line. As to which method is easier in this case, that is up to you to decide.

Figure 7–17 Creating Object by blending

Example 2

What is the best method for creating the U-shaped object shown in Figure 7–18?

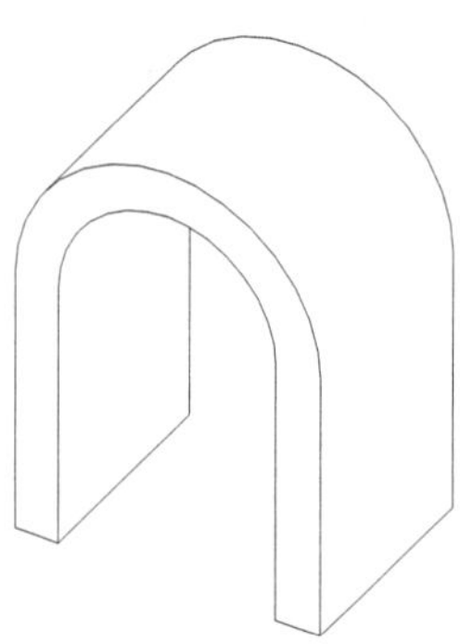

Figure 7–18 Object for Example 2

To create this object by sweeping, you must first create an open section path in the basic U-shape and a closed rectangular cross section as shown in Figure 7–19. You must then sweep the rectangular section along the open path, making sure that it is oriented correctly along the path.

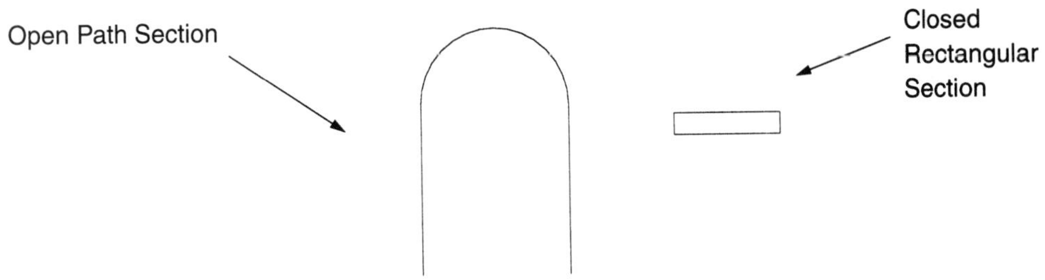

Figure 7–19 Path and Cross Section for Sweeping

A more efficient way to create this object is simply to extrude the cross section shown in Figure 7–20. The time savings realized by this method is that you do not have to worry about how a cross section is oriented along a path. The object created by extrusion will also be easier to modify at a later time than the swept part.

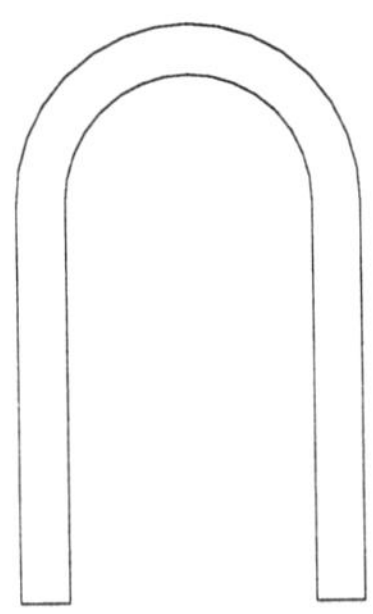

Figure 7–20 Cross Section for Extrusion

Example 3

From the preceding discussion, it should be clear that there are often easier methods for creating objects than blending and sweeping. However, in some cases, blending and sweeping are the best methods for creating objects. For example, what are two methods that can be used to create the rectangular transition piece shown in Figure 7–21?

The easiest method you can use to create this object is to first draw the two rectangular ends as shown in Figure 7–22, then blend these two sections to form the desired object. The other method you could use to create this part is to start with a block and perform four cuts on it to obtain the four tapered sides of the object. Obviously, creating the block and performing the four plane cuts would be a much more complicated and time-consuming operation than creating the two rectangles, moving one, and performing the blend operation.

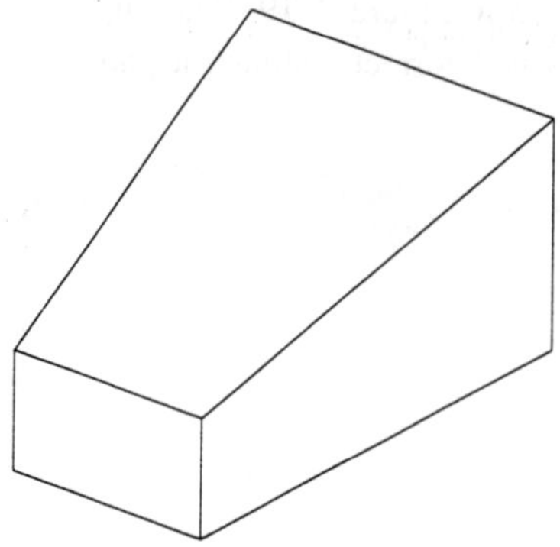

Figure 7–21 Object for Example 3

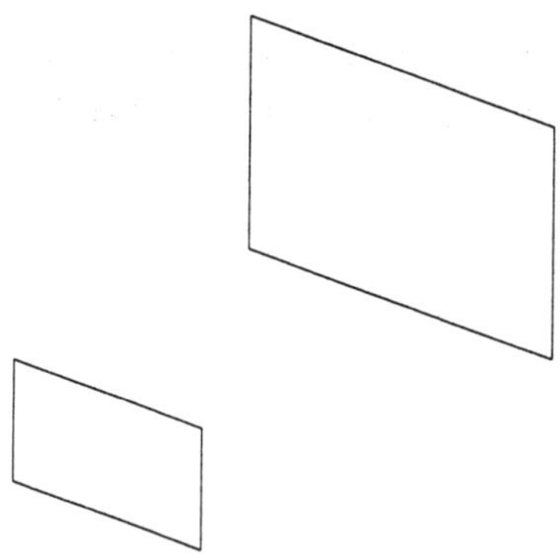

Figure 7–22 Oriented Cross Sections for Blending

Example 4

How would you go about creating the row boat shown in Figure 7–23?

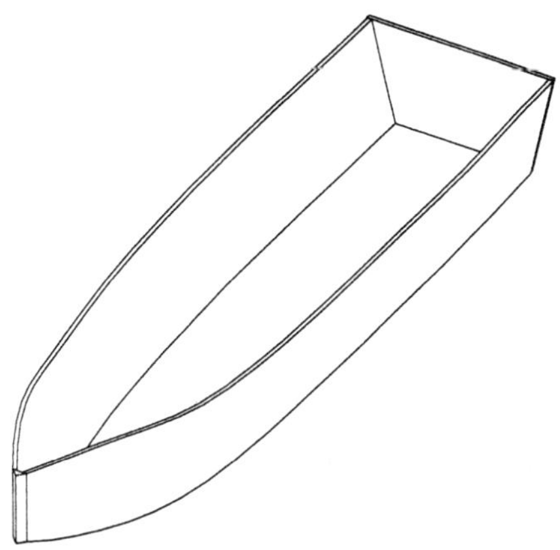

Figure 7–23 Row Boat for Example 4

Blending would work best. Remember to create a closed profile for this—you want the boat to be solid and not just a surface. (Your sketch should be fully constrained, but the constraints and dimensions are not shown here for the sake of clarity.)

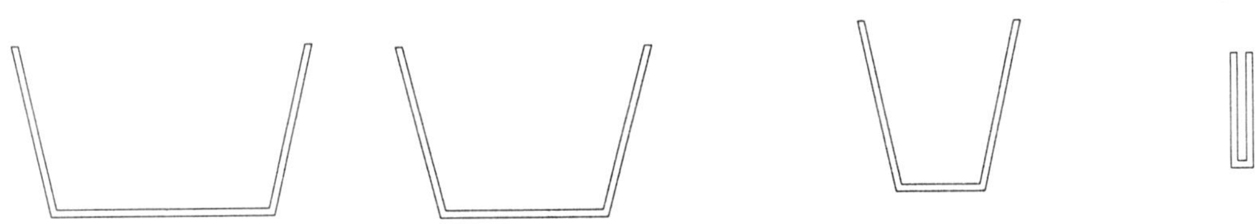

Sketch the cross sections one at a time, and orient them in space to define the main part of the boat. (Make sure you use *Section Tools* and select *Toggle* between the sketching of the cross sections.) The figure here shows the orientation of the cross section from the top, the front, and an isometric viewpoint.

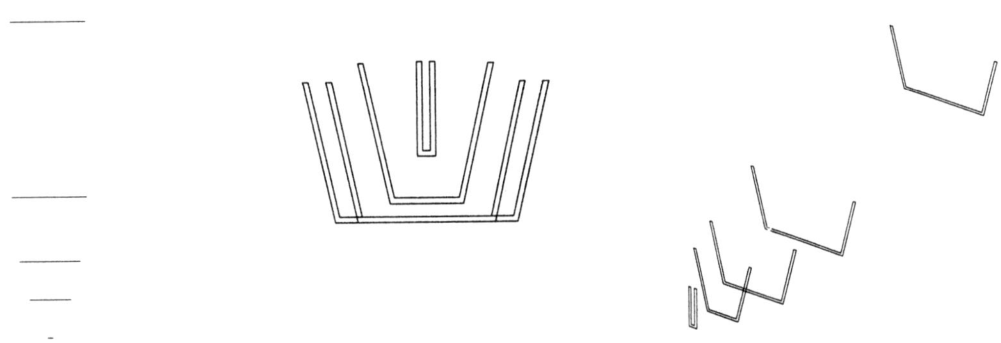

Top View Front View Isometric View

Blend these sections to form the main portion of the boat. When sketching the sections for blending, make sure that you sketch them with a "similar" starting spot (e.g., the upper left corner) for each section to avoid twisting.

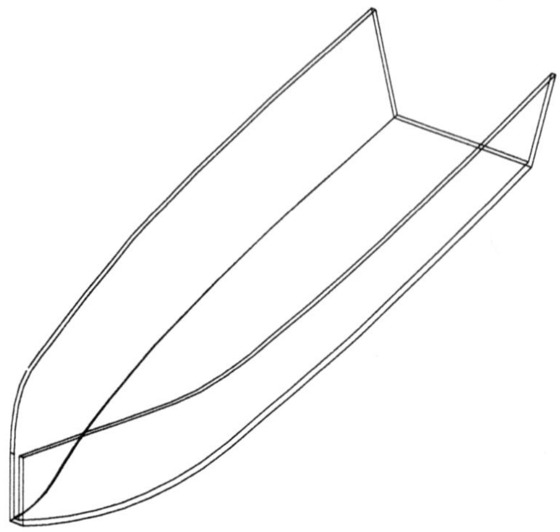

To finish the boat, simply add two extruded features, one on the bow and one on the stern.

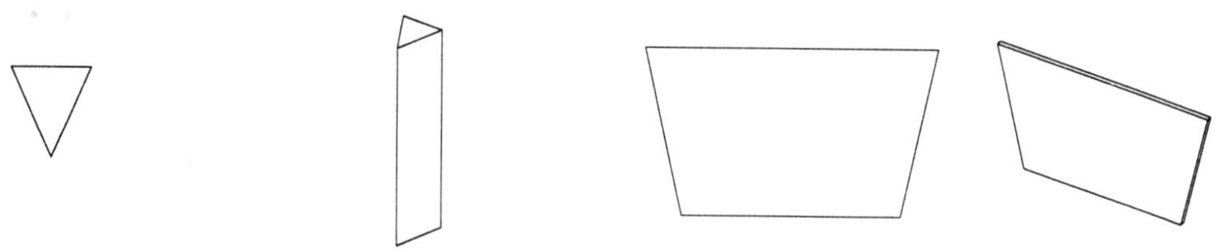

Finished boat should look like this:

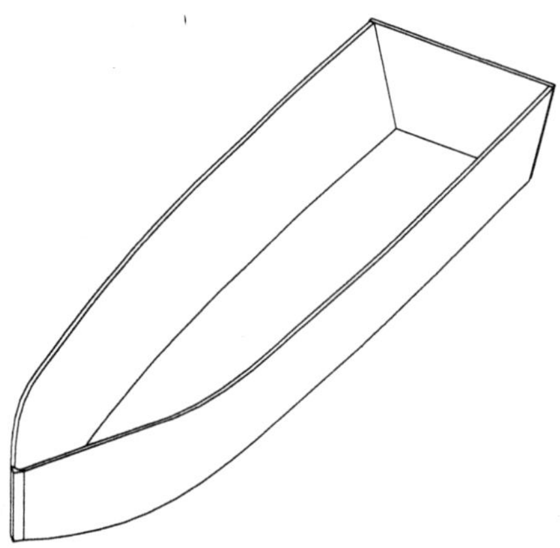

7.8 Troubleshooting/FAQs

Problem	Potential Solution
You are having trouble blending between a circular cross section and a noncircular cross section.	You may want to try dividing the circular cross section up into arcs. Use the *Divide* option (found at the same location as *2-D Fillets*) to do this.
You are blending between cross sections and keep ending up with a "twisted" part.	Note the location and direction of the arrow that Pro/ENGINEER puts on the section. If there is a large variation in the orientation of the arrows between cross sections, you will probably not end up with the blended part you desire. This is especially important when blending between two sections that have circular curves or arcs in them.
You have selected your cross sections for blending, but nothing happened—the form didn't appear so that you could preview a part.	Read the prompts to make sure you have completed inputing the information required by Pro/ENGINEER.
You sketched all the cross sections to blend, but you keep getting multiple loop errors.	Make sure that you have not created intersecting closed loops.

7.9 Chapter Summary

In this chapter you first learned about cross sections and paths and how they can be used to define object geometry. You then learned about blending and sweeping to create complex 3-D objects. The table below list some of the key terms found in this chapter:

Closed Path	Open Path
cross section	Open Surface
Blend	Sweep
Internal Face	Twisted Part

Exercises

On a clean sheet of paper, develop a **Plan of Action** for the object assigned by your instructor describing the steps you would follow to create the object in Pro/ENGINEER. Use sketches as appropriate in your Plan of Action. Object features in the exercises are symmetrical unless otherwise noted. Obtain a view of the object specified by your instructor, and turn in hard copy as appropriate.

1.

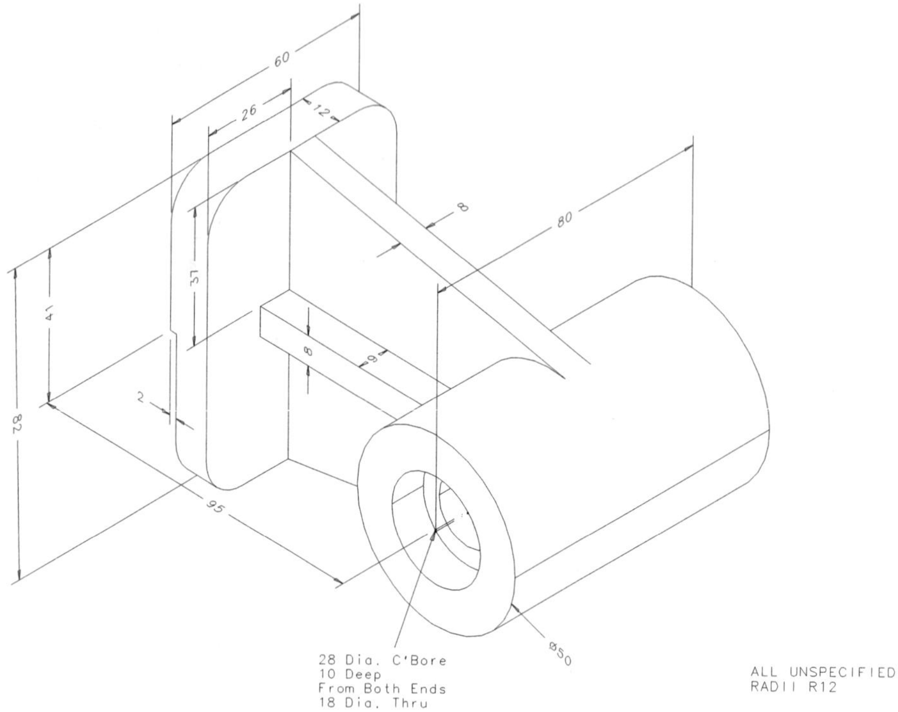

2.

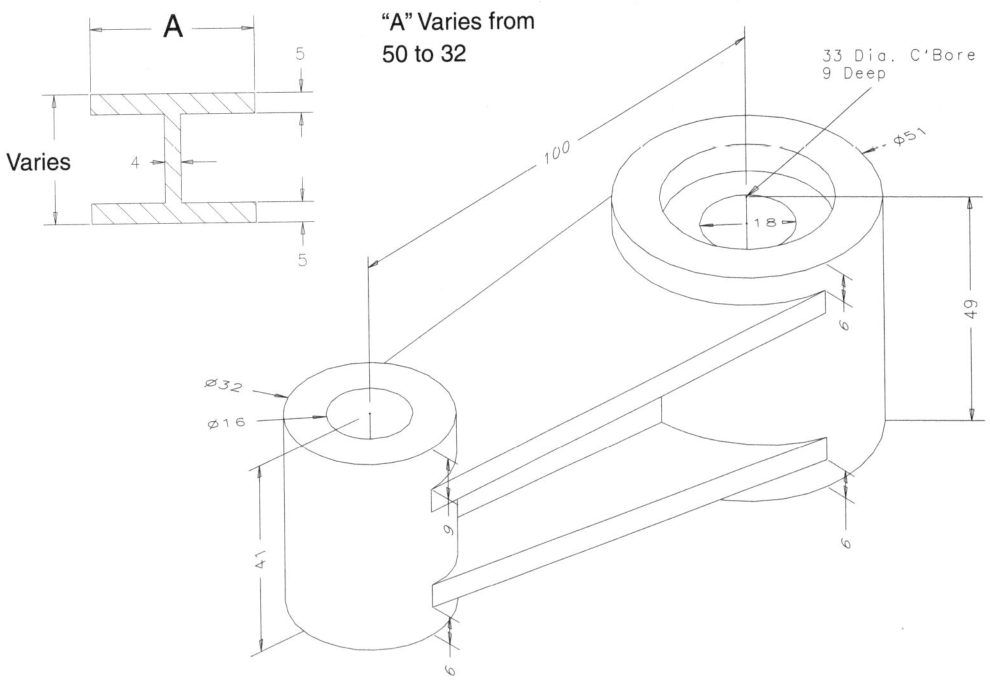
A
Varies
4
5
5
"A" Varies from
50 to 32
100
33 Dia. C'Bore
9 Deep
Ø51
18
49
6
Ø32
Ø16
41
6
6
6

3.

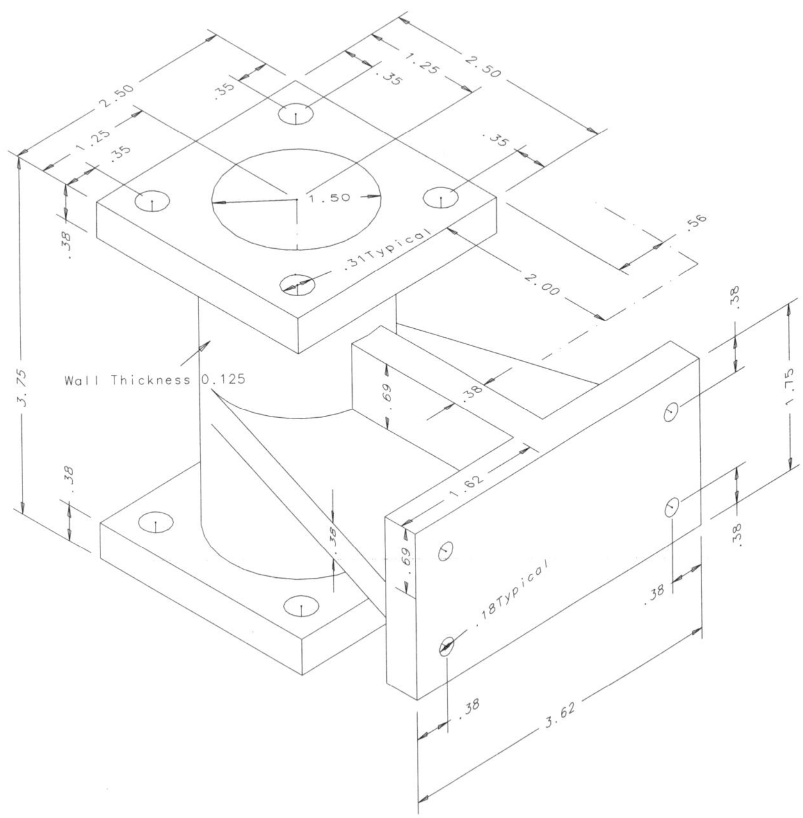
2.50
1.25
.35
2.50
1.25
.35
.35
.35
1.50
.31Typical
.56
2.00
.38
3.75
Wall Thickness 0.125
.69
.38
.38
1.75
1.62
.38
.38
.69
.18Typical
.38
.38
.38
3.62

4.

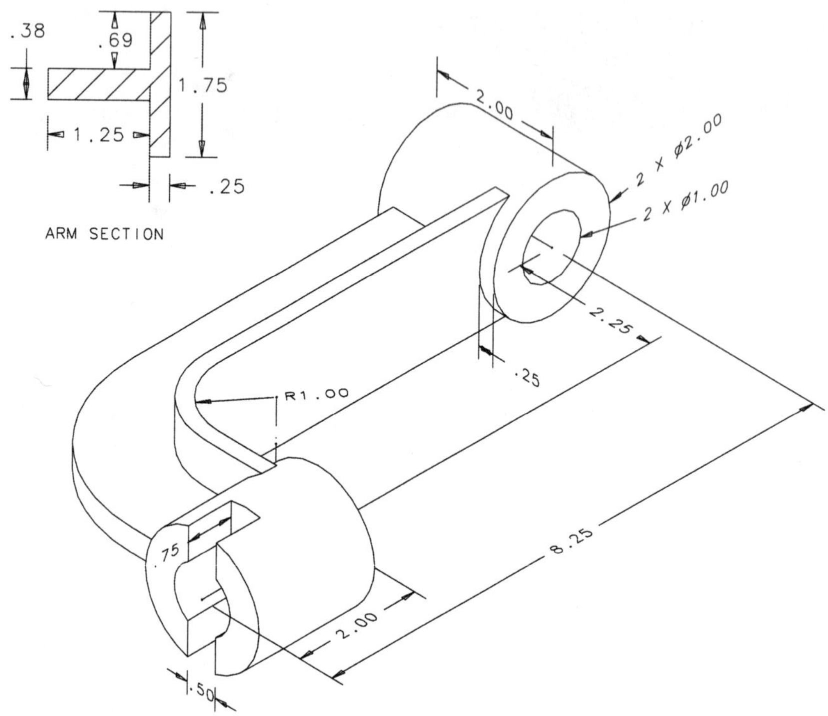
.38
.69
1.75
1.25
.25
ARM SECTION
2.00
2 X ⌀2.00
2 X ⌀1.00
2.25
.25
R1.00
8.25
.75
2.00
.50

Creating an Engineering Drawing

Three-dimensional computer tools have revolutionized the engineering design process. In some cases, machinists can work directly from 3-D computer models in the manufacture of parts or assemblies. In many instances, however, a drawing is still required as the design documentation necessary for manufacture of a part. It is relatively easy to extract a drawing from a three-dimensional solid model, and most of the computer-aided-design software packages that are currently available have this capability. This chapter will cover the creation of engineering drawings from solid models developed in Pro/ENGINEER.

8.1 Creating 2-D Drawings from 3-D Models

In most cases, engineering drawings are drawn to scale due to the fact that it is rarely practical to make drawings that are true size. You are familiar with the concept of a scaled drawing from your experience with maps. Maps are scaled drawings of an area, and the relative scale is usually included in the legend of the map, for example one inch might equal 50 miles. Engineers use scales for drawings that are based on either the metric or the English system of units (inches, feet, etc.).

When creating a drawing from a 3-D model, you first specify the size of the paper on which you wish to create the drawing. In the English system of units, there are five standard drawing sizes: A (8.5" X 11"), B (11" X 17"), C (17" X 22"), D (22" X 34"), and E (34" X 44"). In the metric system of units, there are also five standard drawing sizes that are roughly similar in size to the English paper sizes. The metric standard drawing sizes include: A4 (210 mm X 297 mm), A3 (297 mm X 420 mm), A2 (420 mm X 594 mm), A1 (594 mm X 841 mm), and A0 (841 mm X 1189 mm). The scale you use for your drawing will depend on the size of the part relative to the size of the paper you are using. Typical scales include: true size, half size, quarter size, eighth, or some other scale factor. When setting up your drawing from your 3-D part, make sure that you have enough room on the page for dimensions. The drawing scale does not change the size of the part itself the way you learned about the scale option in Chapter 7 of this text. Specifying a drawing scale only controls your viewing scale of the object (how near or how far you are from the object).

Another option you may need to specify before you create your drawing from a model is whether or not you wish to include a title block on your drawing. If you are working for a company, you will probably include the standard corporate title block on any drawing you create. Otherwise, most software packages have a standard title block you can use in creating drawings.

Engineering drawings typically contain one or more orthographic views of an object and may also contain an isometric or trimetric pictorial for additional clarity. The views that are found on a drawing usually include the top, front, and right side views, although others may be included as necessary. You are already familiar with being able to view a 3-D object from several different viewpoints as you are working with your model in 3-D space. The difference is that you will display the object from multiple viewpoints simultaneously when you are creating a drawing. Figure 8–1 shows a 3-D part and Figure 8–2 shows a four-view drawing that

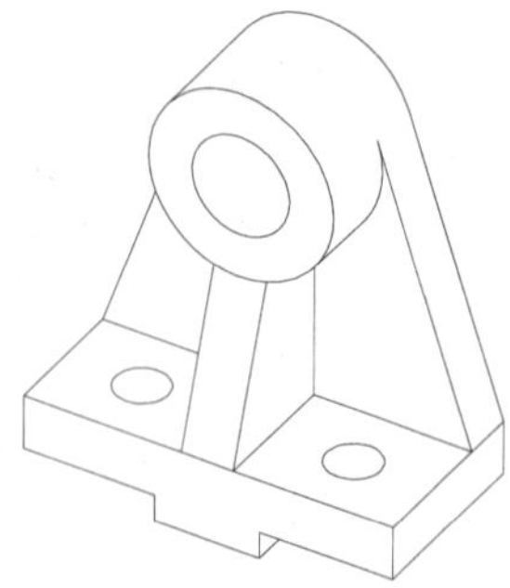

Figure 8–1 Part Geometry

was created from the part. Note that the drawing shows the top, front, right side, and isometric views of the object. Note also that the drawing contains the dimensions in addition to the geometric information.

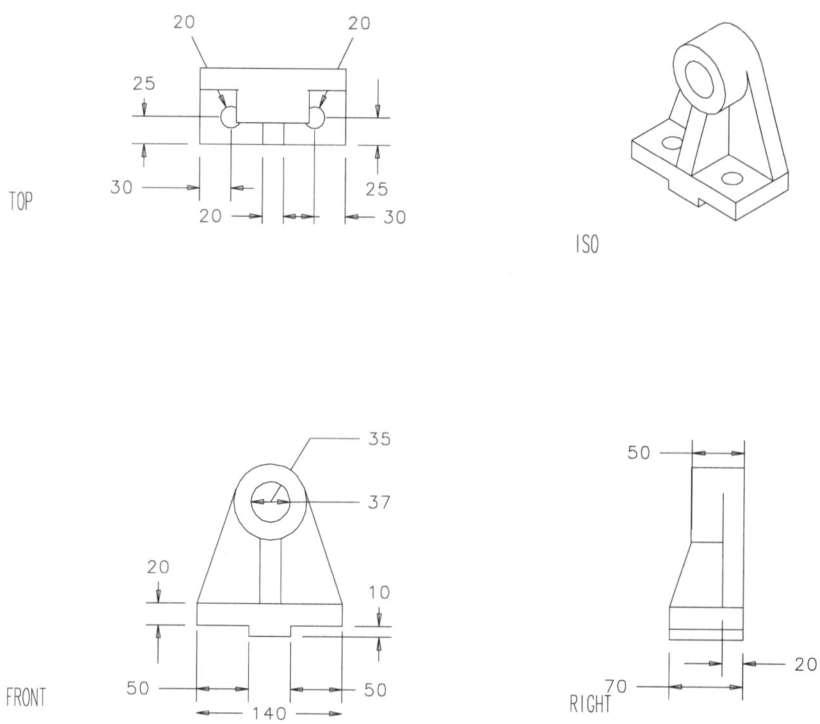

Figure 8–2 Drawing Extracted from Part Geometry

Annotation in the form of dimensions, notes, and labels is commonly found on engineering drawings. Notes are merely text that appears on a drawing, whereas labels consist of text plus an arrow pointing to a specific feature on the drawing. Although dimensions are included as a portion of the object definition, you will often need to edit a drawing generated from a 3-D model to include either more or different dimensions than those that were used to define the part. This type of annotation is easily added.

8.1.1 Special Views

In addition to the necessary views and the drawing annotation, you may need to include additional "special" views common to engineering drawings. These special views typically include details, sections, and auxiliary views.

Detail views are included on a drawing to show a small portion of the part so that it is "zoomed" or enlarged. If you have a feature on a part that is small relative to the overall size of the object, you will probably want to show this feature in a detail view for clarity. Dimensions for the small feature will be added to the detail view rather than the view of the entire part to remove clutter from the drawing.

Sectional views are created by slicing through an object to expose its interior. With these views, numerous hidden lines are not necessary on the drawing itself. Once you expose the interior features, they are no longer hidden. Cross-hatching is included on the drawing where the cutting plane passes through the material. Cross-hatching is considered to be a type of drawing annotation when working with computer-aided-design software tools. There are many types of cross-hatching patterns and many different types of sectional views that can be defined. Because the focus of this text is on solid modeling, these topics are beyond the scope of this text. Figure 8–3 shows a top view and a front sectional view of a piston. Also shown in the figure are a detail view of the region near the location of the grooves for the piston rings and an isometric view of the object.

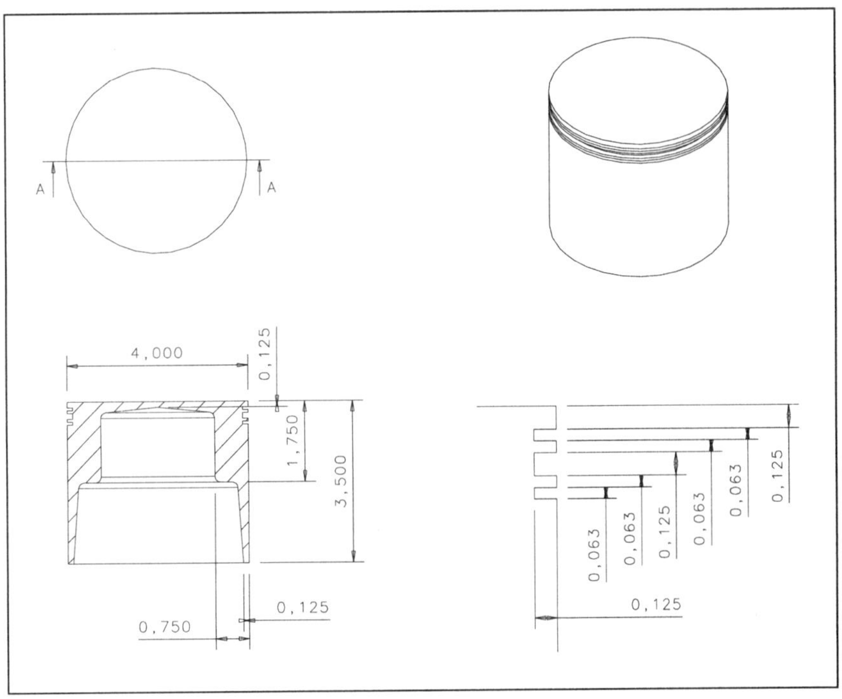

Figure 8–3 2-D Drawing of Piston

Auxiliary views are defined as any view other than one of the six primary views. The primary views consist of the front, top, right side, left side, back, and bottom views. A typical use of an auxiliary view is to show an inclined surface of an object in true-size and shape. This is particularly true if the inclined surface has a feature on it such as a hole that must be located accurately for manufacturing purposes. Since an inclined surface will not appear in true size or shape in any of the primary views, a true-size auxiliary view is necessary to define the location of the hole. Auxiliary views can be constructed to show the true-size view of the inclined surface only, or to show the entire object from this alternate viewpoint, depending on the capabilities of the software. Figure 8–4 shows a drawing of a simple object that contains an inclined surface with holes in it. An auxiliary view showing the location of the holes on the surface is included in the drawing.

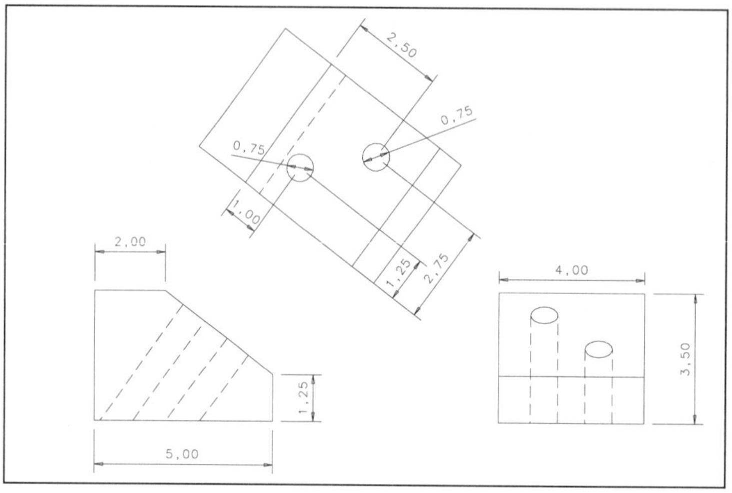

Figure 8–4 Drawing with Auxiliary View

Most modern computer-aided-design tools have either uni- or bidirectional associativity. With unidirectional associativity, if you make a change to the 3-D part definition, any drawings you have created from the part will automatically be updated to reflect the changes made to the model. With bidirectional associativity, if you change the part, the drawing will change, but in addition to this, if you change a dimension on any of the drawings, the 3-D part will change. Associativity is a useful feature of computer-aided-design software in that you do not have to continuously re-create drawings from a part as you modify or change it.

Guided Tour 8.1

- Create the following part. Note that all dimensions are given in inches.
- Print your part in an isometric, no-hidden-line view.

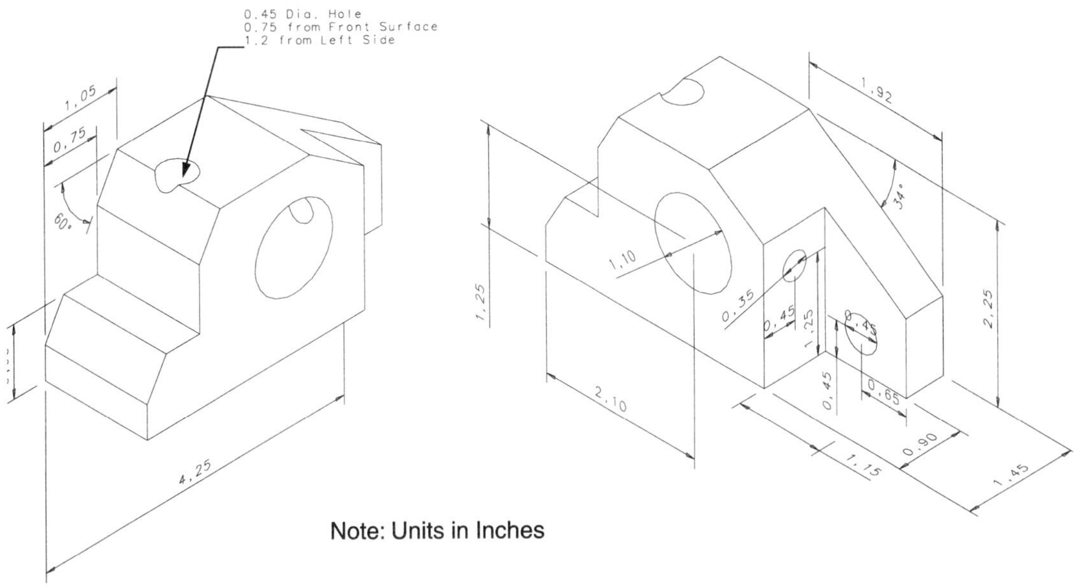

8.1.2 Creating Drawings from Part Geometry in Pro/ENGINEER

Creating an actual engineering drawing from a 3-D part developed in Pro/ENGINEER is a fairly straightforward process. To create the drawing, you first must enter the Drawing mode. This is done by creating a new drawing file. Drawing mode can be used as a drafting tool to create an independent engineering drawing. However, this aspect of the Drawing mode is beyond the focus of this solid-modeling text. The remainder of this chapter will focused on creating a 2-D drawings linked to a previously created solid-models.

To create a 2-D drawing from an object in Pro/ENGINEER, it will be easiest if you first save the parts you are currently working with. From the *File* pulldown menu select *New.* A dialog box will appear to ask you the mode of file you wish to create. Up until now the file mode has primarily been *Part.* Now you will select *Drawing*. After you select *Drawing* and enter the file name of your choice, a new dialog box will appear. In this dialog box, you select the part or model for which you wish to create a drawing, and you also define the drawing size. Pro/ENGINEER will assume that you wish to create a drawing based on the last part you were working with (That model file is listed by default.) Select *Browse* if you wish to use a different model or part file. The paper size is also defined within this dialog box. A complete range of standard paper sizes are available as well as user-defined sizes. From this window, a standard drawing format can be selected. This includes paper size, title block, and border. This is a more advanced feature, and we will return to it later in this chapter.

Once you have finished with the dialog box, Pro/ENGINEER will be in Drawing mode. It is now time to create an engineering drawing. The key purpose of an engineering drawing is to convey your ideas. This is done through the graphic representation of your part. To start this process, select *Views— General—Full View—NoXsec—No Scale—Done*. You will then be asked to select the center point for the first view of your drawing. Since you will most likely be adding other views plan ahead and leave a little extra room. The usual starting point for this first view is in the lower left quadrant of the drawing area. Click the left mouse button in the center of this quadrant. A view of your part should now appear on the screen. Pro/ENGINEER will prompt you to orient this part. It is important that you take the time to orient your part now as it governs the placement and alignment of subsequent views.

Guided Tour 8.2 - View Layout and Orientation

- Start a new drawing by selecting *File—New.*
- In the dialog box which appears verify or enter the name of your new part from Guided Tour 8.1.
- Select *Views— General—Full View—NoXsec—No Scale—Done.*
- Click the left mouse button in the lower left quadrant of the screen. Your screen should look similar to the following figure:

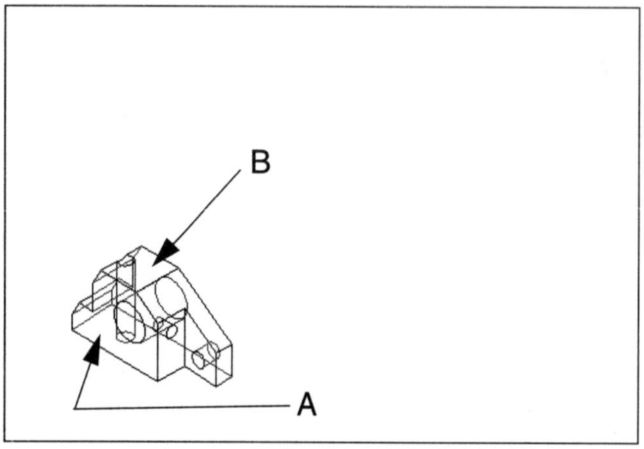

- At the prompt to orient the part, *query select* the face "A" for the front, reference 1, and face "B" for the top, reference 2. Pro/ENGINEER should orient your part so that it appears flat against the screen. Select *OK.* This view will be referred to as the **front view**.
- Select *Add View—Projection—Full View—No Xsec—No Scale—Done*
- Click a few inches above the front view. A new view, the top view, should appear.
- Select *Add View—Projection—Full View—No Xsec—No Scale—Done*

- Click a few inches to the right of the front view to create a right-side view. Your drawing should look similar to the next figure.

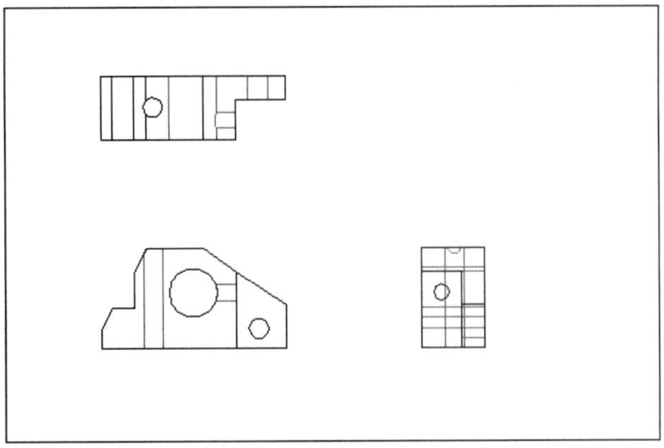

- To add an isometric or trimetric view select: *Add View—General—Full View—No Xsec—No Scale—Done*. When prompted to orient your part, do not simply select *OK*. You should end up with a drawing similar to the following figure:

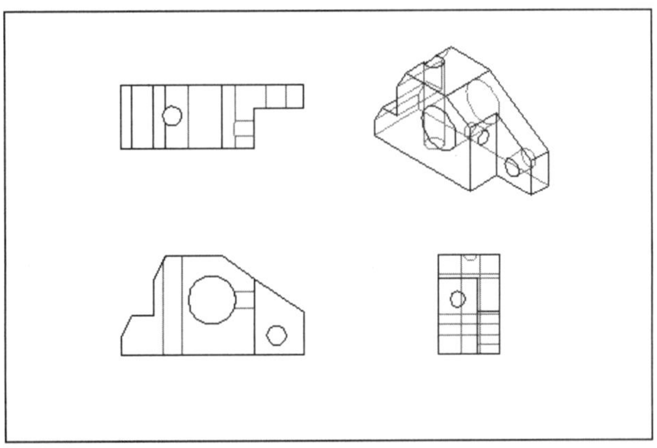

- Print out your results.

- Select *Move View* and click on the front view. What happens?

Once you have created the layout, the next step is to add the dimensional information. To add dimensions to your drawing, select *Detail—Show/Erase—Show*, verify that the dimension button is activated (the upper left most button), and select *Show All—Close*. This should generate a mess of dimensions, most of which will be unreadable. Now, the task is to clean them up. It is important to note that the dimensions shown here are the ones you used to create the model. These dimensions are called "Key Dimensions," and they cannot be deleted from the drawing since they are a part of the object definition. Key Dimensions can only be removed from the drawing by use of the **Erase** function in *Show/Erase*. The Key Dimensions will still exist, but they will remain hidden.

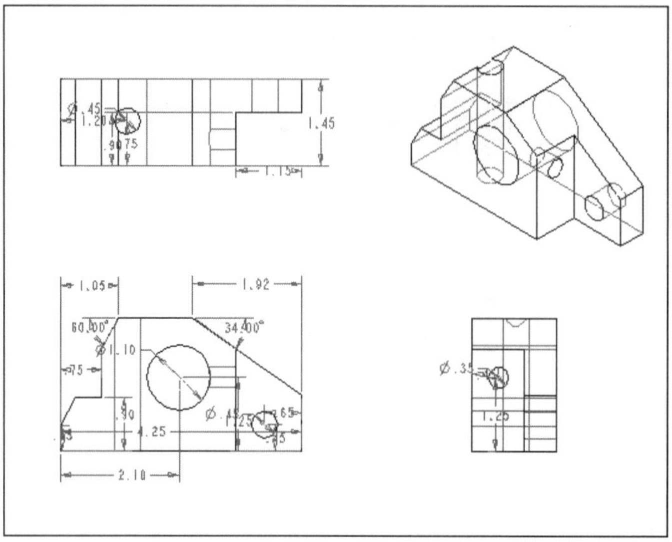

To clean up the dimensions select *Move* from under *Detail* in the Menu Manager. This is may appear to be a tedious task, but it is an important one. If your dimensions are hard to read, your part may be built incorrectly. An improved drawing follows:

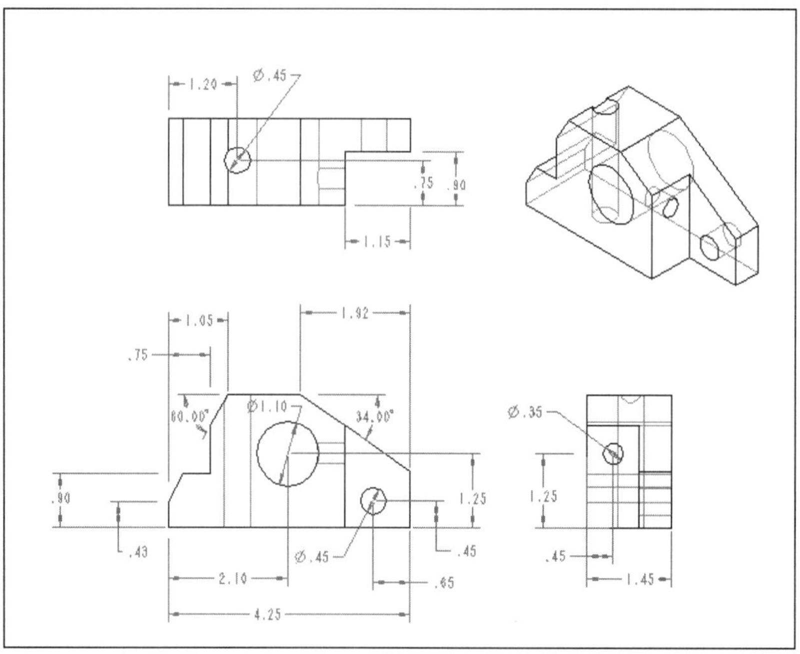

8.1.3 Adding Views to Drawings

Once you have created a basic drawing from the part, you must go about the job of cleaning it up. You may wish to include additional views on the drawing. Two additional views that are commonly found on engineering drawings are the detail view and the sectional view. This section explains what these views are and how they are added.

Detail View. The detail view is a tool engineers and designers use to highlight small and intricate areas on a part. A detail is added using *Views—Add View—Detail—Full View—No Xsec—Scale*. Simply follow the instructions in the message window to place this new view. Most often you will use a scale factor greater than 1.00. For a step-by-step procedure, see Guided Tour 8.3.

Guided Tour 8.3 - Detail View

- Start with the drawing developed in Guided Tour 8.2, and add the dimensions as discussed in the text.

- Select *Views—Add View—Detail—Full View—No Xsec—Scale*

- You should now be prompted to choose the location of the new detail view. Select an area slightly to the upper left from the front view.

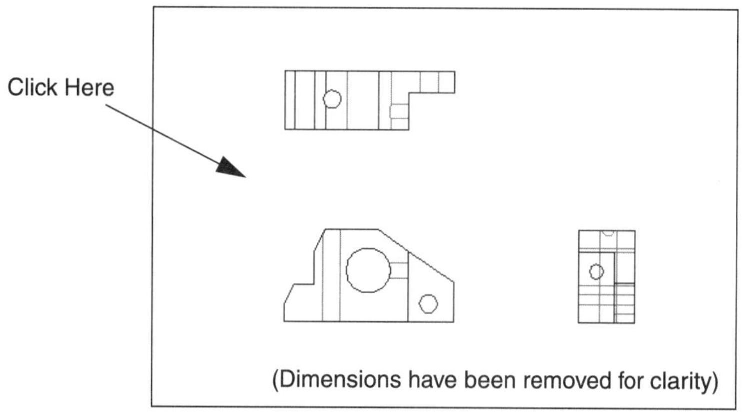

- Enter a scale factor of 2.0.

- You should now be prompted to select the "center point for detail on existing drawing." This is asking you to select the center of the area you wish to "zoom in" on. Select a point close to the one indicated in the following figure:

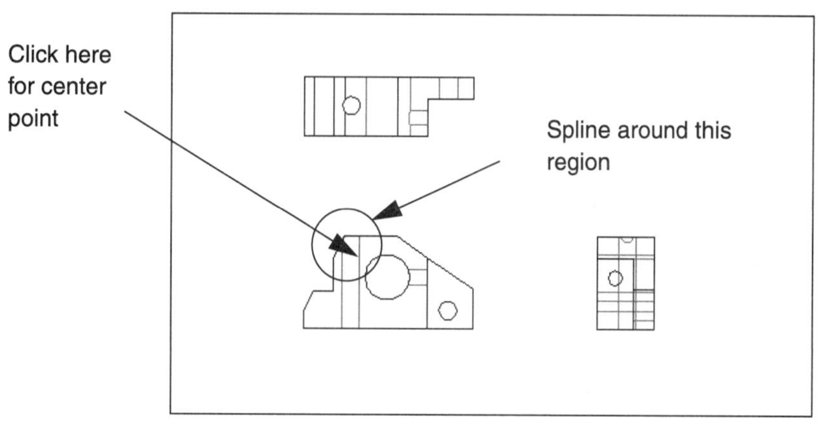

- Using the left mouse button and the spline, encircle the area of interest. The more left-mouse-button clicks you make, the smoother your spline will be. Use the middle mouse button to close the spline.

- At the prompt for the detail name, type *A*.

- Select the boundary type (usually a circle).

- Finally, indicate the location of the note. Your drawing should look similar to the following

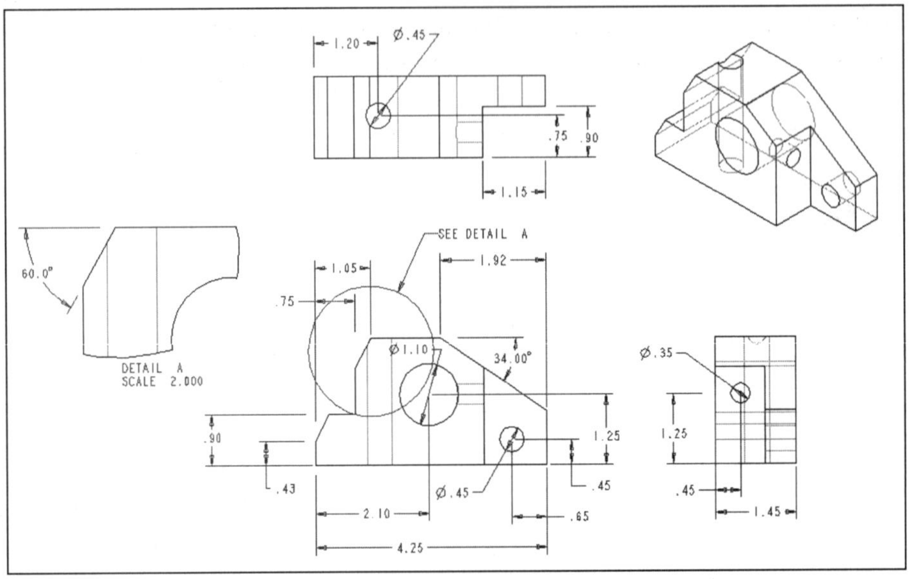

Section View. The section view is a fairly advance drawing tool. In order to simplify the creation of the section view, it is best to go back to the original model in Part mode and create a datum plane where you wish to cut the part (Refer to Chapter 6, if necessary). To begin the process of adding a section view, you must select *Views—Add View—Projection—Full View—Section—No Scale—Done*. Following that sequence of commands, you need to define the type of cross section. In most cases it will be *Xsec Type—Full—Total Xsec—Done*. You will then be prompted for the location of this new view. The second part of creating section views is defining the cutting plane. The sequence of required commands is *Create—Planar—Single—Done*. Enter the name of the section view (usually just one letter) and then, either *Query Select* or select the name of the datum you created as the cutting plane. The following Guided Tour will step you through this process.

Guided Tour 8.4 Section View

- Return to Part mode (File *Open* will work) and create a datum plane 0.75 inch from the front surface.

- Return to Drawing mode (under the Window pulldown menu, toggle through the windows, and activate the drawing started in Guided Tour 8.3).

- Select: *View—Add View—Projection—Full View—Section—No Scale—Done*

- Next, select *Xsec Type—Full—Total Xsec—Done*

- You should then be asked to indicate the location of the new view. Select a point to the right of the right-most view.

- Then select *Create—Planar—Single—Done.*

- At the prompt for the section name, enter *B.*

- At this point, Pro/ENGINEER needs you to specify the cutting plane. By taking the time earlier to create a new datum plane, this process was greatly simplified. Use either *Query Select* or *Select by Name*. If you *Select by Name*, pick *Datum* and then follow the instructions in the message window.

- Your Drawing should look similar to the following one:

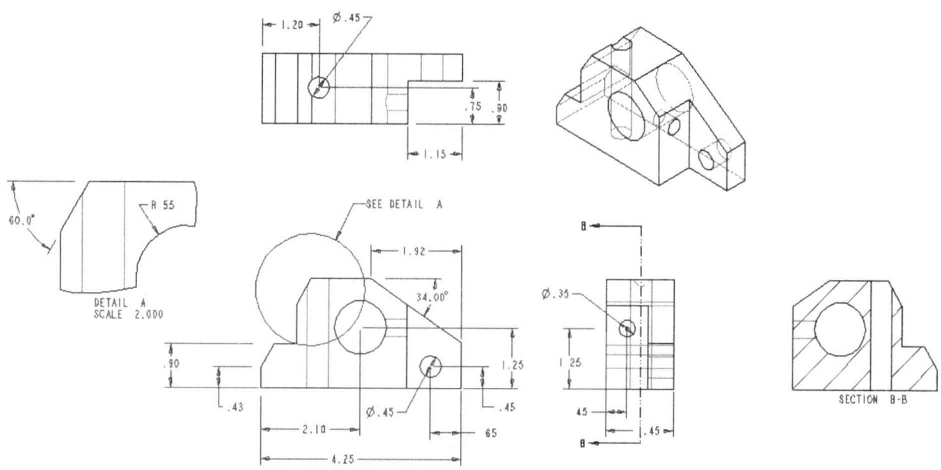

8.1.4 Modifying Drawing Views

You may wish to modify the display of views on your drawing by moving, deleting, or changing the scale factor associated with the views.The options you will use to change the views on your drawing are outlined in the following:

Move. Use this option to move a view on the drawing to a new location. When moving a view, a dashed line will appear around the view to be moved. If other views were introduced as projections of this view, all views will be moved to maintain proper orthographic projection. The command sequence to move a view is *Views—Move View—Pick.* Select the view you wish to move.

Delete. Use *View—Delete View* to completely remove a view from the drawing. Care must be taken so you do not delete a view upon which other views are based. For example, don't try to delete the front view if the top and right sides are projections of that view.

Scale. Changing the size or scale of one of the views is simple using Pro/ENGINEER. The command sequence is *Views—Modify View—Change Scale—Pick.* Select the view for which you wish to change the scale. Careful attention must be made when selecting the view because only a view which was created with a scale can be changed.

Perhaps the most useful tool for modifying the views in Pro/ENGINEER is the use of the **Display Modes** feature. Display Modes allows you to select, by view, various parameters to suit your tastes. The use of Display Mode is helpful to remove the hidden lines from the isometric and section views as are commonly found on other engineering drawings.

Guided Tour 8.5 - Display Modes

- Starting from Guided Tour 8.4, select *Views—Display Mode—View Disp.*
- When prompted, select the isometric view.
- Now select *No Hidden—Done.*
- Work through the same procedure on the section view. When you redisplay your screen, your drawing should look similar to the following

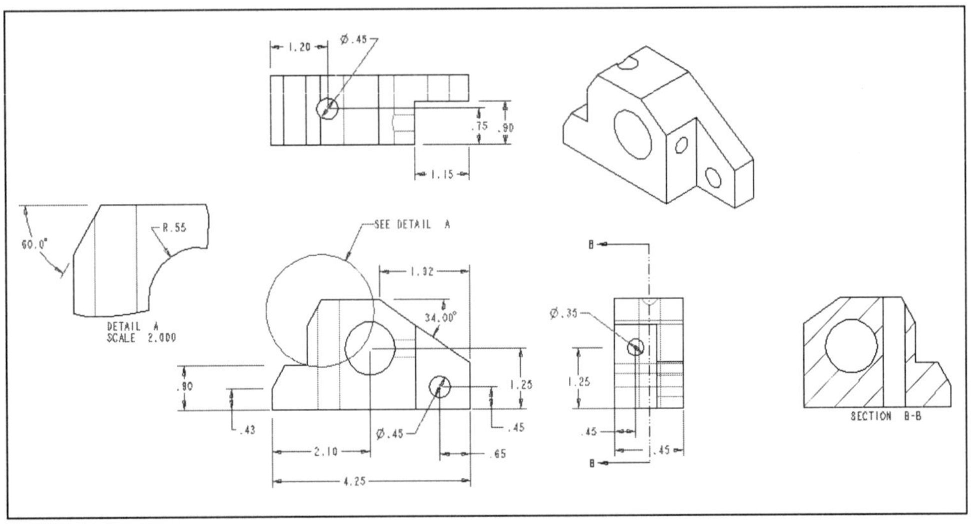

8.1.5 Adding/Changing Drawing Annotation

Annotation is a very important part of an engineering drawing. For this section, we will focus the discussion on adding additional dimensions, reference dimensions, and notes.

Until now, the only dimensions on your drawing were the Key Dimensions. The Key Dimensions were created automatically as you created your part, In addition to these dimensions, you have the ability to include user-defined dimensions that you add to the drawing to help clarify your design intent. If you modify these dimensions, there will be limited associativity with the part, since they are not part of the object definition but are only placed on the drawing itself. The command sequence to add user-defined dimensions is as follows: *Detail—Create—Dimension.* Then simply select points, lines, planes, etc. The method of placing the dimension is very similar to that of adding a dimension in Part mode: left mouse button to select the first point, left mouse button to select the second point, and middle mouse button to place the dimension. If Pro/ENGINEER is unable to determine your intended orientation of the dimension, it will ask you to select horizontal, vertical, slanted, parallel, or normal. If you don't understand the differences in these options, experiment with them and see what happens.

Another key annotation feature is the use of reference dimensions. It is common practice to avoid putting duplicate dimensions, which describe a given feature on the part. However, it can be more convenient if we add a few extra dimensions to help make the part easier to build. These duplicate dimensions are know as reference dimensions. To add a reference dimension, select: *Detail—Create—Ref Dim*. Then, much like the user- defined dimensions, select the different features to be measured. The difference will be noticeable once the middle mouse button is pushed, indicating the location for the dimension. The value for the dimension will be shown with REF behind it. This is a note to the person reading the drawing that a more descriptive dimension exists elsewhere on the drawing for that given feature. Some general dimensioning rules to keep in mind as you go about creating drawings are as follows:

- The various dimension features should be created so that they do not interfere, obscure, or coincide with object, hidden, or center lines on the drawing.
- Crossing of extension lines and dimension lines should be avoided if possible.
- Dimensions should reference object lines rather than hidden lines.
- Do not repeat dimensions on the same or another view.
- Dimension and extension lines should plainly indicate their point of application. However, there should be a visible "gap" between the end of the extension line and the object feature it is referencing.

Another consideration in adding dimensions to your drawing is that you generally want your dimensions to be included from edge to edge and not from point to point on the drawing. When dimensioning your part, you will probably need to keep the display of the center lines and the center points turned on so that you have selectable points to assist you in dimensioning.

Adding plain text is done through the command sequence *Detail—Create—Note*. This allows you to write simple text on the drawing, call out specific machining processes, and add drawing or part names. The *Create Note* command is useful when filling in the title and revision blocks (discussed in a following section). Other annotations such as geometric tolerancing and surface finish are found under the *Detail—Create* command but are beyond this text.

In addition to adding dimensions and text and to moving them around on your drawing, you may need to change the appearance of the dimensions on the drawing to suit your particular needs. For example, you can choose to place the arrows either inside or outside the witness lines; you can set the color of the dimension, the size of the dimension text, set the units or the number of decimal places that are displayed; you can change the size or the style of the arrowheads; you can control the size of the extension lines and leaders; and you can control the gaps associated with these entities. The options for customization are numerous. Modifications are made by selecting *Detail—Modify*. To modify the style of text, select *Text—Text Style*. If you deselect height and width factors, you can really have some fun. Try it and see.

8.1.6 Adding a Title Block and Format

When you first created your drawing, you specified many options on the original form. Some of the options were the model to create a drawing of, the paper size, and format selection. Until this point, we have been creating drawing without a format. Pro/ENGINEER refers to the format as the title block, revision block, and borders (see the next figure). An engineering drawing is not truly complete without the format added and filled in correctly. The format contains the title of the drawing, the company name, the date last revised, the part number, the material specifications, and the designer's name. This information is very useful in design management.

To add a format to a drawing that has already been created, select: *Drawing—Sheets—Format—Add—Replace*. A dialog box should appear, and you will need to change to the format directory. (Click on the little folder icon near the top of the dialog box.) Once you are in the format directory, select the format for the appropriate size work space you are using. For must student applications, this will be "a.frm". To add text, use the *Detail—Create—Note* commands.

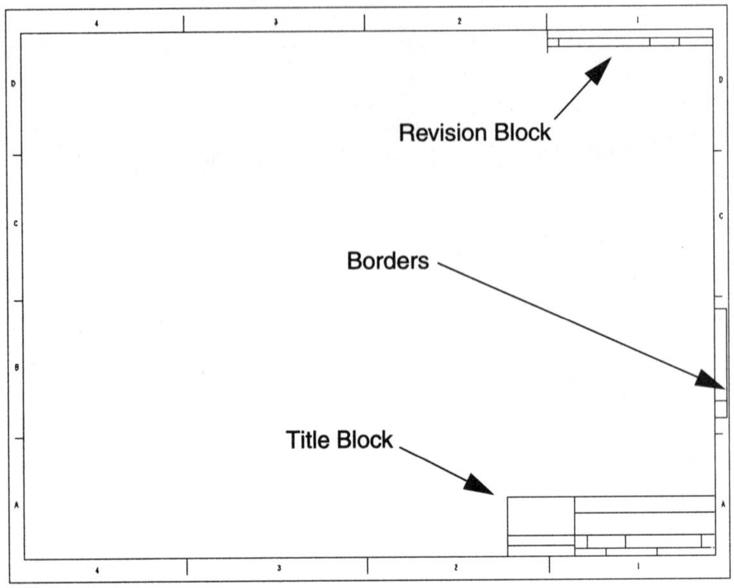

Guided Tour 8.6

- Add a format to the drawing from Guided Tour 8.5.
- Add your name to the title block. When finished, your drawing should look similar to the next drawing:

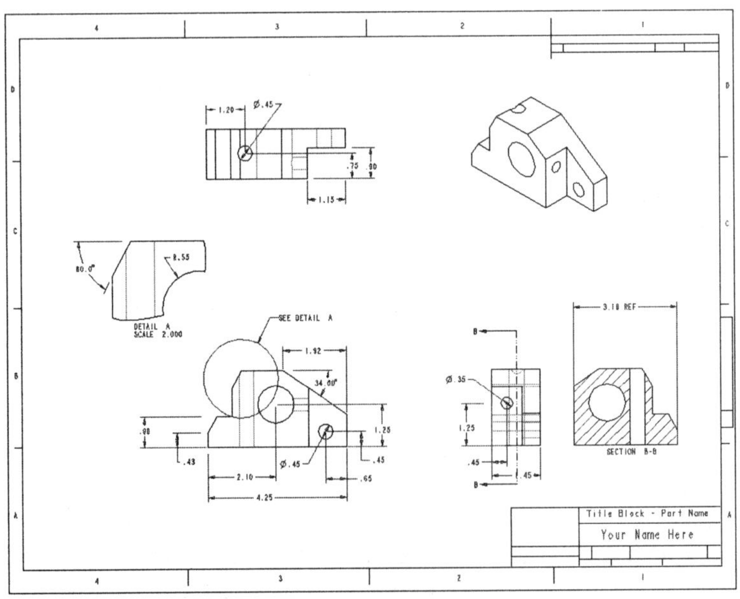

- Change the hidden line style to *dashed.* Add your name as a note on the drawing (without a leader line). Obtain a printout of your drawing.

8.1.7 Managing Your Drawings

As stated previously, Pro/ENGINEER has associativity between parts and drawings. Once the drawing is created, it is attached to the part definition. For this reason, if you have created a drawing from a part, you must not delete the part itself from your Pro/ENGINEER directory (because the drawing will no longer be able to make reference to the part). You can also create multiple, different drawings from a single part. Thus, a single part could have two or more drawings attached to it.

8.2 Two-Dimensional Drawing Strategy

One of the main things you should consider in creating drawings from parts is determining the number of views required to completely describe the object. Some objects need only two views to describe them, whereas other objects may require three orthographic views as well as sectional, detail, and auxiliary views. The inclusion of an isometric view is not generally required, but it can be included for complex parts for clarity.

Example 1

Figure 8–5 shows an isometric view of an object. How many views do you think are required to completely describe this object?

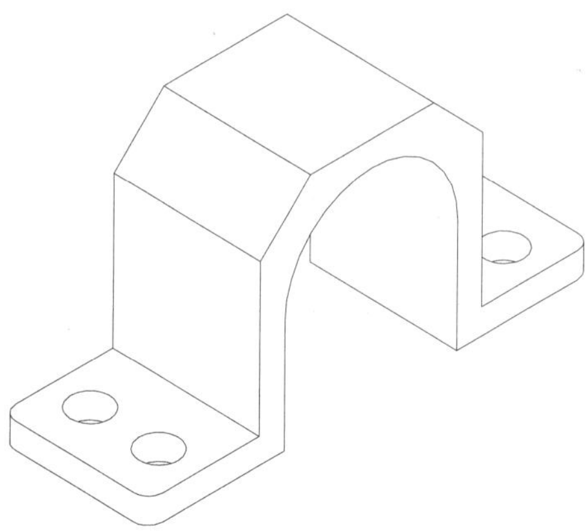

Figure 8–5 Object for Example 1

For this particular object, only two views are required to completely describe it as shown in Figure 8–6. The general shape of the object is visible in the front view, and the locations of features such as the small holes are visible in the top view. If the right side view were included for this object, no new information would be gleaned. Extraneous views should be avoided in clear graphical communication associated with engineering drawings.

Example 2

How many views are required for the object shown in Figure 8–7?

Three views are required to completely describe this object as shown in Figure 8–8. If the right side view were eliminated, the size of the hole, and the corner locations for the inclined surface could not be established. Likewise, if the top view were eliminated, the shape of the filleted corners would not be obvious, and the two holes could not be sized or located accurately. The front view is also essential for describing the object because it shows the semicircular feature as well as the filleted corners.

Another consideration in setting up your drawing is deciding which view should be the front view of the drawing. In general, the most descriptive view of the object should be placed on the drawing as the front view. In general, the most descriptive view is the one that most accurately portrays the general shape of the object.

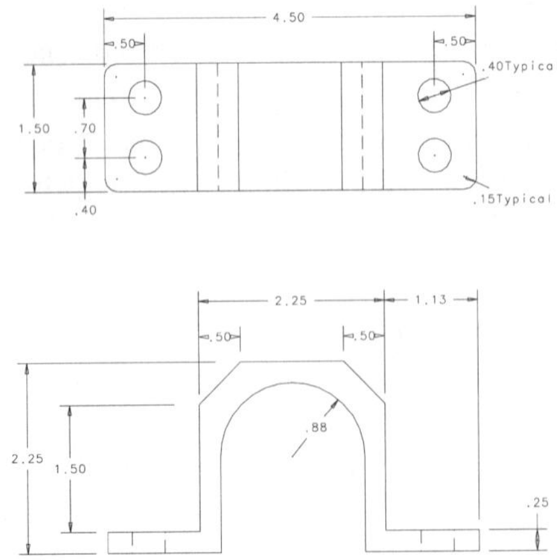

Figure 8–6 Drawing of Object for Example 1

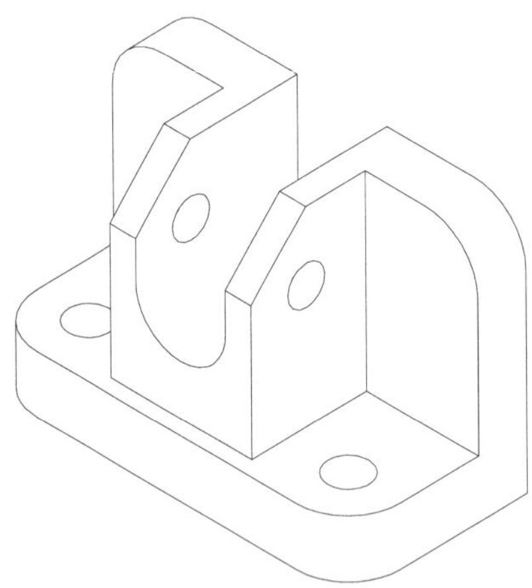

Figure 8–7 Object for Example 2

Please note that Figure 8–9 is not drawn completely to conventional practices. There are circle centers missing, the projected hole in the side view is not dimensioned according to convention, etc. The complete strategy that you employ to create 2-D drawings is beyond the scope of this text. Proper creation of 2-D drawings is covered in many excellent engineering graphics texts. When creating drawings from 3-D parts using a computer, the most important thing to remember is to plan carefully and to allow enough room for dimensions or other annotation to be added later. However, unlike hand drafting, if you make a mistake in the initial setup of your drawing, it can be easily fixed without starting over. Careful planning will avoid time-consuming modifications later on.

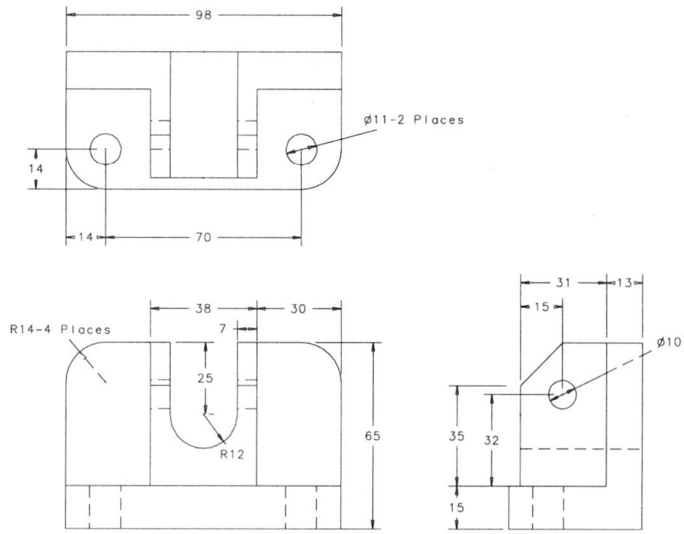

Figure 8–8 Object Drawing for Example 2

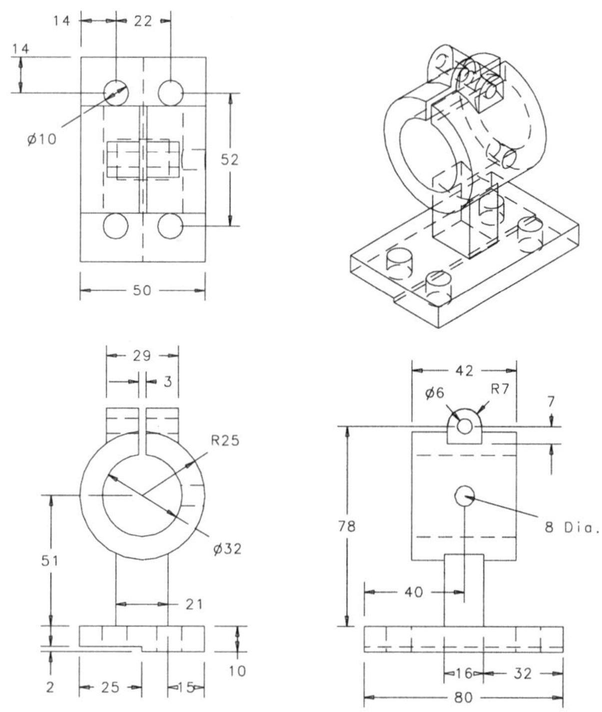

Figure 8–9 Completed Drawing

8.3 Troubleshooting/FAQs

Problem	Potential Solution
The projected views of your object don't line up.	Chances are you created these views incorrectly by bringing each view in as "general." Only the first view and the isometric view should be brought in as general, the rest should be brought in as "projection." A projected view will not need to be aligned.
You want to dimension to the center of a circle but can't pick that location for dimensioning.	Make sure you are using *Query Select.*
You want to delete a dimension from your drawing, but Pro/ENGINEER won't let you.	Chances are you are trying to delete a Key Dimension. Key Dimensions will appear in cyan, and these are the dimensions you used in the creation of the part. Key Dimensions can *not* be deleted, but you can use the Hide icon to turn off their visibility.
The dimensions are there but don't make much sense and certainly don't follow any conventional practices.	Use *Show/Erase* to remove unwanted Key Dimensions, then use *Create* to specify user-defined dimensions.
The dimensions on your drawing do not contain a visible "gap" between the extension lines and the object geometry, or the gap is too big.	Select *Move* from the Menu Manager and click on the extension line. You will be able to move its endpoint.
You decide that you want to change the size of the paper for your drawing.	Select: *Drawing—Sheets—Format—Add/Replace.* Then specify the new size format.

8.4 Chapter Summary

In this chapter you learned how to create a 2-D drawing from a 3-D part. Sectional, detail, and auxiliary views were defined as well as drawing annotation. Conventional practice for modifying drawings created from parts was also covered in this chapter. Key Words for the chapter are tabulated in the following:

Add View	Key Dimensions
Annotation	Material Type
Associativity	Move (View)
Auxiliary View	Note
Cross-hatching	Open Surface Area
Cross Section	Properties
Detail View	Scale Views
Dimension	Section
Drawing Scale	Sectional Views

Exercises

On a clean sheet of paper, develop a **Plan of Action** for the object assigned by your instructor, describing the steps you would follow to create the object in Pro/ENGINEER. Then, create the object in Pro/ENGINEER making sure that it is the correct *size* as well as the correct *shape*. Object features in the exercises are symmetrical unless otherwise noted. Create a drawing layout for the part, including those views on the drawing as assigned by your instructor. Note to the Instructor: You could assign a drawing setup for any previously assigned object at this point in time

1.

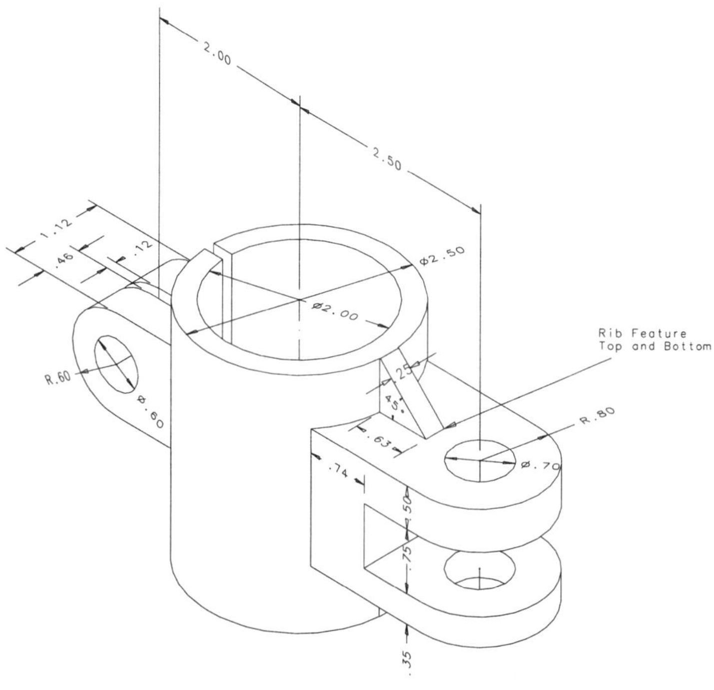

2.

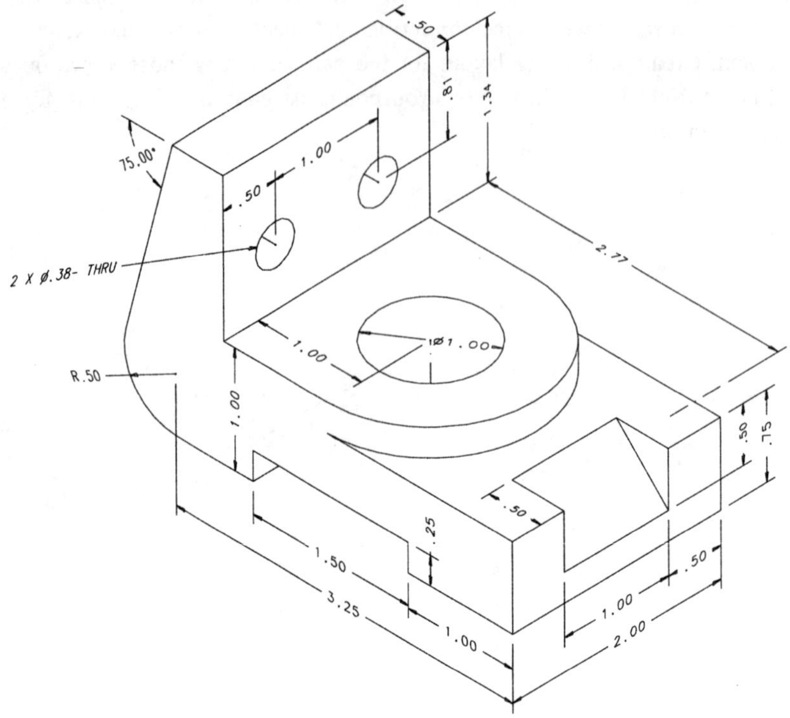
.50
.81
1.34
75.00°
1.00
.50
2 X Ø.38- THRU
2.77
R.50
1.00
Ø1.00
1.00
.50
.75
.50
.25
1.50
3.25
1.00
.50
1.00
2.00

3.

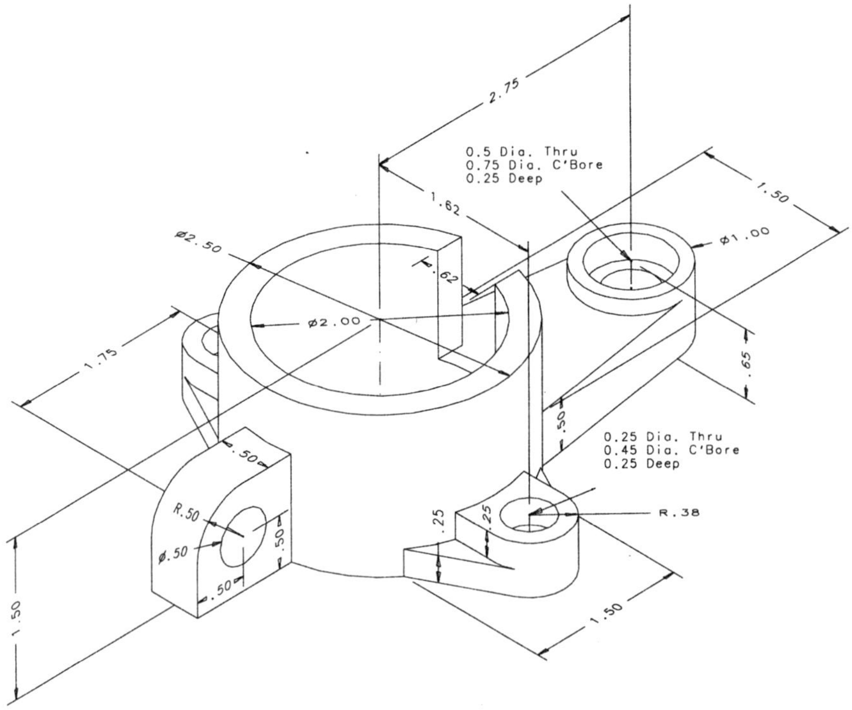
2.75
0.5 Dia. Thru
0.75 Dia. C'Bore
0.25 Deep
1.50
1.62
Ø2.50
.62
Ø1.00
Ø2.00
1.75
.65
.50
0.25 Dia. Thru
0.45 Dia. C'Bore
0.25 Deep
.50
R.50
Ø.50
.50
.50
.25
.25
R.38
1.50
1.50

4.

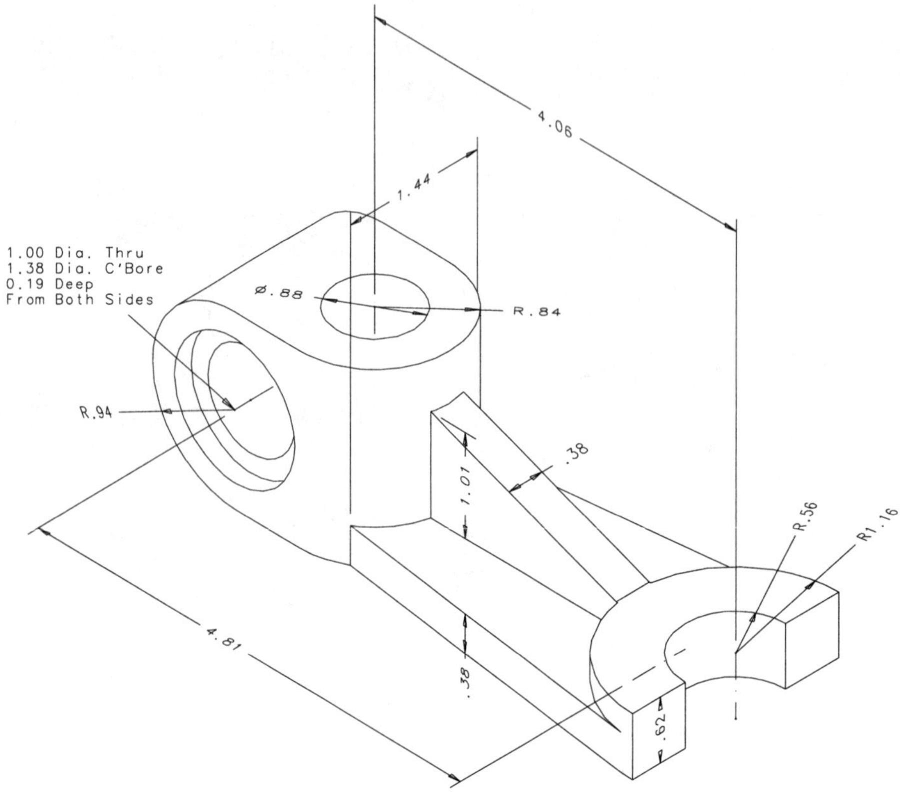
1.00 Dia. Thru
1.38 Dia. C'Bore
0.19 Deep
From Both Sides
4.06
1.44
Ø.88
R.84
R.94
.38
1.01
R.56
R1.16
4.81
.38
.62

Index[1]

[1] Italics designate commands; boldface pagenumbers designate figures; page numbers followed by "t" designate tables